WAXCAP MUSHROOMS OF EASTERN NORTH AMERICA

WAXCAP MUSHROOMS
of Eastern North America

· ALAN E. BESSETTE · WILLIAM C. ROODY ·
· WALTER E. STURGEON · ARLEEN R. BESSETTE ·

SU
SYRACUSE UNIVERSITY PRESS

NOTICE: Although *Waxcap Mushrooms of Eastern North America* contains information regarding the edibility of various mushrooms, this book is **not** intended to serve as a guide for the safe consumption of wild mushrooms. Before eating any wild mushrooms, readers should consult other sources of information such as experienced mycophagists, mycologists, and guidebooks that more specifically address these matters. Neither the publisher nor the authors are responsible for any undesirable outcomes that may occur for those who fail to heed this warning.

Syracuse, New York 13244–5290

First Edition 2012
12 13 14 15 16 17   6 5 4 3 2 1

∞ The paper used in this publication meets the minimum requirements of the American National Standard for Information Sciences—Permanence of Paper for Printed Library Materials, ANSI Z39.48-1992.

For a listing of books published and distributed by Syracuse University Press, visit our Web site at SyracuseUniversityPress.syr.edu.

ISBN 978-0-8156-3268-9

LIBRARY OF CONGRESS CATALOGING-IN-PUBLICATION DATA

Waxcap mushrooms of eastern North America / Alan E. Bessette ... [et al.]. — 1st ed.
p. cm.
Includes bibliographical references and index.
ISBN 978-0-8156-3268-9 (cloth : alk. paper) 1. Mushrooms—East (U.S.)—Identification.
2. Mushrooms—Canada, Eastern—Identification. I. Bessette, Alan.
QK605.5.E27W38 2012
579.60974—dc23   2011050392

Title spread: *Hygrophorus speciosus* var. *speciosu*

Book design by Christopher Kuntze
Printed and bound in Canada by Friesens

Contents

Preface

Members of the Hygrophoraceae family, commonly known as "waxcaps," have long attracted the attention of mycologists and nature lovers. As a group, they are beautiful mushrooms, and those in the genus *Hygrocybe* are particularly colorful and eye catching. Adding to their appeal for those lacking formal training in mycology is that many waxcaps can be identified from field observations and macroscopic features of the fruiting bodies. However, some species can be identified only with the aid of a microscope and even then with considerable difficulty.

Waxcaps are usually well represented in general mushroom field guides, and there have also been more thorough scientific treatments of the North American species. Excellent as these previous works may be, they have not satisfied the desire for a nontechnical yet comprehensive guide that illustrates the eastern North American waxcaps in color. It has been more than forty years since the publication of L. R. Hesler and A. H. Smith's classic monograph *North American Species of Hygrophorus* (1963). Although this book still remains the principal work on which all subsequent North American studies (including the present work) are based, it did not include color illustrations, and several additional species have been described since its publication. The work presented here is not intended to be a scientific treatment of the Hygrophoraceae, but we believe it will fill a gap between sporadic coverage in general mushroom field guides and the more inclusive technical monographs.

As with most groups of fungi, the systematics in Hygrophoraceae have undergone review in recent years. This review has resulted in differing opinions regarding classification schemes. Indeed, some mycologists have abandoned the concept of the family Hygrophoraceae altogether and placed *Hygrophorus* and *Hygrocybe* in the Tricholomataceae family. The recent merging of the genus *Camarophyllus* with *Hygrocybe* also has its detractors. It is not our intention to favor one view over another in these matters, but for the ease and convenience of our readers we have chosen to retain the family Hygrophoraceae and place the species we include here into the two principal genera *Hygrocybe* and *Hygrophorus*. This arrangement is the most easily comprehended for field studies.

The geographical range of coverage includes eastern Canada, the United States south to Texas, the Gulf Coast, and Florida west to the Rocky Mountains. Although the distribution of species is constantly being expanded as knowledge accumulates, most waxcaps that occur within this region are featured or discussed. No taxonomic

OPPOSITE: *Hygrocybe cantharellus* f. *cantharellus*

work is ever truly complete because new species are sure to be discovered either in nature or as a result of modern molecular studies. We have purposely omitted a few waxcaps that have rarely been recorded for the region and therefore not likely to be encountered.

We offer this book as an illustrated reference and guide for the identification of these common and attractive mushrooms. We hope the selected photographs do some justice in showing the aesthetic beauty of waxcaps as they appear in nature. The descriptions are based on the original authors' concepts and on subsequent modifications from our own and others' observations.

Acknowledgments

Our grateful appreciation goes to many individuals who have contributed in various ways to this work. We thank Sam Norris for the skillful line drawings of gill trama types that are shown in the introduction. D. Jean Lodge provided taxonomic notes and technical assistance. Donna Mitchell reviewed the draft manuscript and made several helpful suggestions. We are indebted to David Largent, Renée Lebeuf, David Lewis, John Plischke III, and Michael Wood, who contributed photographs that significantly enhanced the book's beauty and functionality. Our gratitude also to Seanna Annis, Ike Forester, Hope Miller, and Steve Stephenson for kindly allowing us to include photographs in their possession from the collections of the late Richard Homola, Dan Guravich, Orson K. Miller Jr., and Emily Johnson, respectively. We especially appreciate the efforts and contributions of Annie Barva, who copyedited the manuscript, and Christopher Kuntze, who designed the book. Finally, this book would not have been possible without the support of Alice Randel Pfeiffer and the staff at Syracuse University Press. Thank you all.

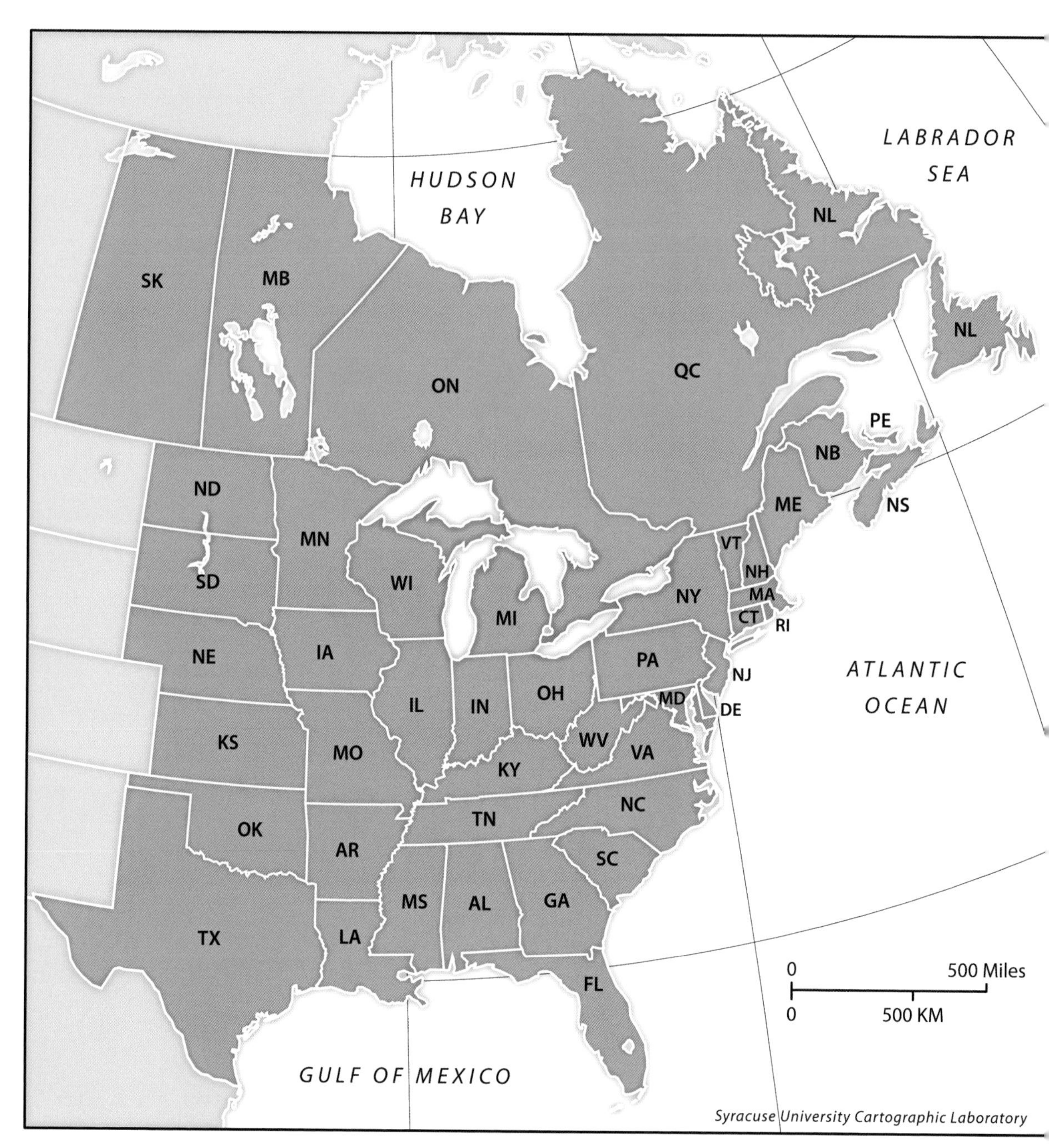

Distribution range map for waxcap mushrooms of eastern North America

WAXCAP MUSHROOMS OF EASTERN NORTH AMERICA

Introduction

WAXCAPS DEFINED

Waxcaps are fairly distinctive mushrooms that can be readily recognized after one becomes generally acquainted with the group. The most obvious macroscopic feature that distinguishes waxcaps from other white-spored gilled mushrooms is the translucent luster of the gills, which have a clean, waxy appearance. Many have a perceptible waxy "feel" that is remarkably similar to soft candle wax when the cap tissue and especially the gills are rubbed or crushed between the fingers. The gills are thick but taper to a sharp knife-like edge and are more or less triangular in cross-section. On most species, the gills are moderately to widely spaced.

Microscopic measurements indicate that the waxcaps' basidia are at least five times longer than the length of the spores. The elongated basidia give the gills their characteristic waxy appearance. The waxy feel may be attributed to the chemical structure of the basidial wall or to the turgor pressure of the basidial or hyphal contents (Young 2005). The waxy quality is not apparent on dried specimens.

The arrangement of the hyphae of the gill trama is the principal microscopic feature for distinguishing the genera of waxcaps. In *Hygrocybe*, the gill trama hyphae are parallel or interwoven, whereas in *Hygrophorus* they are divergent (bilateral) (see the glossary for a further explanation of these terms).

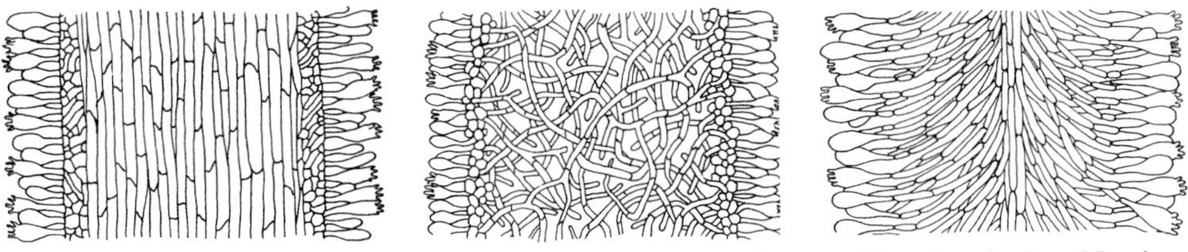

l to r: Parallel gill trama; Interwoven gill trama; Divergent gill trama. Drawings by Sam Norris.

Waxcap spores are ellipsoid to nearly cylindrical or globose. They are smooth or rarely slightly roughened. With rare exceptions, the spores are inamyloid. Under the microscope, the spores are colorless, but accumulated in mass they produce a white to cream spore print. Although variation in spore ornamentation is lacking, spore size can be useful in determining some species. An in-

OPPOSITE: *Hygrocybe conica* var. *conica*

depth detailed consideration of microscopic features can be found in Hesler and Smith's 1963 monograph and in other technical literature (see Resources and Recommended Reading).

A BRIEF OVERVIEW OF CONTRIBUTORS TO THE KNOWLEDGE OF WAXCAPS IN EASTERN NORTH AMERICA

Charles Horton Peck (1833–1917) was the first mycologist to produce a monograph on American waxcaps. As botanist for the state of New York, Peck began to describe and catalog the fungi of New York in a series of annual reports from 1868 until his retirement in 1915. He was astonishingly prolific and was the first to describe more than twenty-five hundred species of larger fungi. Although his interest was broad, Peck had a special affinity for the agarics. His 1906 monograph on New York *Hygrophorus* included forty-one species, of which fourteen were either new at the time or previously described by him.

The next American mycologist to make major contributions to the study of waxcaps was William Alphonso Murrill (1869–1957). Like Peck, Murrill was a tireless worker. While associated with the New York Botanical Garden from 1904 until he retired in 1924, he described hundreds of species of fungi. Later in life, Murrill moved to Florida and became associated with the University of Florida in Gainesville, where he continued to describe more than six hundred new species of mushrooms, including southern and subtropical species of *Hygrophorus.* In his earlier writings, following the Finnish mycologist Petter A. Karsten (1834–1917), Murrill used the name *Hydrocybe* instead of *Hygrophorus.* During his illustrious career, Murrill was the first to describe thirty-three species of waxcaps.

It was not until 1963 that the first scientific monograph covering all of the known North American waxcap species was written by Lexemuel R. Hesler (1888–1977) and Alexander H. Smith (1904–1986). Smith and Hesler had previously collaborated on two major taxonomic articles on American *Hygrophorus* in the journal *Lloydia* (1939 and 1942). This work was eventually superseded by the publication of their monograph *North American Species of Hygrophorus* (1963), which featured 244 species and varieties. Hesler was a prominent mycologist whose studies centered on the larger fungi of the southern Appalachians and southeastern United States. Smith, at the University of Michigan, was a giant in American mycology whose productivity was on a par with Peck and Murrill. In addition to their work on *Hygrophorus,* Hesler and Smith jointly authored a number of important monographs of the agaricales, including *Lactarius, Pholiota,* and *Crepidotus.*

In 1979, C. J. Bird and Darryl W. Grund coauthored "Nova Scotian Species of *Hygrophorus,*" which featured fifty-three species and varieties, including two previously undescribed species. They closely followed the taxonomic structure set forth by Hesler and Smith. This regional monograph contains both technical and field keys as well as black-and-white photo illustrations. It remains a very useful resource for the study of waxcaps throughout eastern Canada and the northeastern United States.

In 1985, Richard L. Homola (1934–1999), while a mycologist at the University of Maine, along with Miroslaw M. Czapowskyj and Barton M. Blum, authored a nineteen-page booklet titled *The Ectomycorrhizae of Maine 3: A Listing of Hygrophorus with the Associated Hosts,* in which forty-four species of waxcaps are shown with color photos. Although this small publication does not include species descriptions, it serves as a natural compliment to Bird and Grund's monograph.

Also in 1985, David Largent of Humboldt State University authored an unbound monograph on *Hygrophorus* as part of the Agaricales of California series, with an optional supplement of seventy-eight color photos. Although its emphasis is on the occurrence and distribution of species within California, it includes several cosmopolitan waxcaps that can also be found in eastern North America. With few exceptions, the photos are studio depictions of collected specimens.

Other American mycologists who were the first to describe one or more species of eastern waxcaps include George F. Atkinson (1854–1918), Howard E. Bigelow (1923–1987), William G. Cibula (1932–2005), Moses A. Curtis (1808–1872) in collaboration with British mycologist Miles J. Berkeley (1803–1889), Charles Frost (1805–1880), Charles H. Kauffman (1869–1931), Louis C. C. Krieger (1873–1940), Orson K. Miller Jr. (1930–2006), Andrew P. Morgan (1836–1907), Louis David von Schweinitz (1780–1830), and Nancy S. Weber. D. Jean Lodge and Sharon Cantrell are currently studying the hygrocybes of the Lesser Antilles, and it is likely that some of these tropical species may eventually be recorded from southern Florida and the Gulf Coast.

There have been numerous taxonomic treatments of waxcaps in Europe, and several species found in North America were first described by European mycologists. Because many of the same species occur on both continents, these works are surprisingly useful for American researchers, especially those written by Eef Arnolds and David Boertmann (for the latter, see Resources and Recommended Reading). Works by Rolf Singer (1906–1994), Peter D. Orton (1916–2005), and Roy Watling are also recommended for those who have an interest in waxcaps on a global scale. Our apologies to those past and present whom we may have inadvertently overlooked in this condensed history.

THE GENUS *HYGROCYBE*: GENERAL MACROSCOPIC FEATURES

For the most part, species of *Hygrocybe* are small, shiny, brightly colored mushrooms that have attached adnate to adnexed or free gills. Many are some shade of yellow, orange, red, or white, but a few are gray to brown or nearly black. The recent merging of the genus *Camarophyllus* into *Hygrocybe* has broadened the concept to include some species that are typically dull colored and not shiny and that have decurrent gills. None of the hygrocybes has a universal or partial veil.

Cap

The cap shape ranges from conic to convex, convex with an umbo, or nearly flat to slightly funnel-shaped. The surface is glabrous to somewhat scaly or fibrillose. It may be moist or dry or viscid and slippery. Some caps are translucent-striate on the margin. Cap color is not an entirely reliable feature because pigments may fade with age or wash out in wet weather. The flesh is solid, lacks a distinctive odor in most species, and can be mild or bitter tasting. Few are known to be edible.

Gills

The gills are typically thick and distant to subdistant and most often attached to the stalk, although some are free or become free (seceding) as the cap expands.

Stalk

Hygrocybe stalks are thin, straight, equal or compressed, solid or hollow. The surface is smooth, moist or dry, or viscid and slippery. In some species such as *Hygrocybe psittacina* and *H. irrigata,* the stalk is so slippery that it is nearly impossible to grasp. Even stalks that are initially dry may become lubricous when handled.

Ecology

Habitat and Substrate: Hygrocybes can be found singly or in groups on soil or humus (infrequently on well-decayed wood and moss-covered logs) in forests, grasslands, grassy woodland margins and clearings, and among mosses. Some species occur commonly in bogs and sand dunes. *Hygrocybe spadicea* var. *spadicea* and *H. acutoconica var. cuspidata* are frequently found on the ground in burned areas.

Season: Late spring through fall or winter in the far South.

Ecological Role: Species of *Hygrocybe* are presumed to be saprobes, but not all relationships are fully understood. For instance, *Hygrocybe andersonii* fruits only in the rhizosphere of the shrub Florida Rosemary *(Ceratiola ericoides).* This pattern outwardly resembles an ectomycorrhizal fruiting, but if so, it would be unusual for *Hygrocybe*. It is more probable that this species is saprobic on the shrub's root system. Several hygrocybes grow in or among mosses. Whether the association with various bryophytes is coincidental or perhaps something more specific is an area that needs further study.

THE GENUS *HYGROPHORUS*: GENERAL MACROSCOPIC FEATURES

Members of the genus *Hygrophorus* are typically medium to large mushrooms (occasionally with caps up to 20 centimeters wide). Many are dull or dark-colored, whereas others are pallid or white. Some have a glutinous universal veil, which causes the cap and stalk to be slimy when fresh. Debris firmly affixed to the cap cuticle of dry or drying specimens and a shiny coating on the lower stalk are indications of the presence of a glutinous veil.

Cap

Hygrophorus caps are convex to obtusely conic at first but may become flattened or funnel-shaped with age. An umbo is present on some. The surface is glabrous to fibrillose or scaly. Some are moist or dry, but many are viscid or glutinous and shiny. The flesh is thick and solid, without a distinctive odor in most species. However, when pronounced, odor can be an important diagnostic character. Although some caps are bitter tasting, several have mild flesh and are edible.

Gills

The gills are typically thick and often subdistant to distant, but a few species such as *Hygrophorus russula* and *H. bakerensis* have close to crowded gills. They are broadly attached to the stalk (adnate) or to varying degrees descend down the stalk (decurrent). Except when debris-spattered from heavy rain, the gills are very clean looking. In some species, they become stained or spotted in age.

Stalk

The stalks of *Hygrophorus* are central, solid, and moderately thick for most species. The surface is smooth and glabrous or may be scurfy or granulose-dotted, especially at the apex. The stalks are moist or dry or viscid to glutinous. A glutinous universal veil, which is present on some species, is not always obvious because the gluten may wash off in rainy weather or disappear in dry conditions. A few species exhibit a fibrillose annular zone from a remnant nonglutinous veil.

Ecology

Habitat and Substrate: Species of *Hygrophorus* occur singly or in groups on soil or humus near trees and shrubs. All species of *Hygrophorus* are thought to form ectomycorrhizal associations with various trees and shrubs, especially with oaks and pines.

Season: The vast majority of *Hygrophorus* species appear from summer to late fall or throughout the winter months in the Deep South. Several are adapted to cold conditions and often fruit after the first freeze.

EDIBILITY OF WAXCAPS

Waxcaps do not receive much respect when it comes to their culinary attributes. Although very few species are known or suspected to be poisonous, only a small number is regularly collected for eating. The waxy texture and bland flavor do not endear waxcaps to connoisseurs of edible mushrooms. The thick glutinous veil on some species of *Hygrophorus* also makes them unappealing to collect and difficult to clean. However, these qualities vary considerably from species to species, and in fact some waxcaps are rather good edibles. Charles McIlvaine, who was perhaps the most enthusiastic wild mushroom eater of all time, said in referring to *Hygrophorus virgineus* and *H. niveus* in his classic *One Thousand American Fungi* (1900), "In the West Virginia mountains, along grass-grown roadsides, their purity and exquisite perfume attracted me in 1881. I have them and a few others to thank for seducing me into becoming a mycophagist" (154).

When discussing the genus *Hygrophorus* in *The Agaricaceae of Michigan* (1918), C. H. Kauffman opined that "most are to be classed among our best edible mushrooms" (174). In *Common Edible and Poisonous Mushrooms of Southeastern Michigan* (1938), even Alexander H. Smith, who admittedly was more of a scientist than mycophagist, characterized many *Hygrophorus* species as being "exceptionally fine food" (30).

Waxcaps seem to have lost favor more recently with eaters of wild mushrooms. In *Mushrooms Demystified* (1986), David Arora says of waxcaps, "I have yet to find one to my liking. By and large they are too bland or too watery or too bland *and* too watery to be worth eating" (104). In our opinion, the truth lies somewhere between these opposing views. We have found *Hygrophorus subsordidus* and *Hygrocybe pratensis* (especially var. *pallida*) to be quite good and well worth collecting for the table. *Hygrophorus flavodiscus,* although not particularly flavorful, is mild tasting and has the advantage of fruiting late in the season when few other edible mushrooms are available. *Hygrophorus camarophyllus* is reported to be very good. Some collectors apparently enjoy *Hygrophorus russula,* although all our collections of this species have been bitter and unpalatable. We suspect that other waxcaps may prove to be acceptable or desirable edibles if put to the test, but they are often dismissed or passed over in favor of better-known edible wild mushrooms.

oster announcing a mushroom fair n Cataluna, Spain, with waxcaps as its heme

Waxcaps for sale at the famous La Boqueria market in Barcelona, Spain

In other countries where wild mushrooms are routinely gathered for food, waxcaps are better appreciated. Several species are commonly sold in the mushroom markets of Europe. In the French and Spanish Pyrenees, *Hygrophorus latitibundus* (= *Hygrophorus limacinus*) is a highly regarded edible that is even commercially canned for sale in supermarkets. Other waxcaps that are popular edibles in Europe include *Hygrophorus dichros, H. gliocyclus, H. marzuolus, H. penarius, H. poetarum,* and *H. russula*. Well before the first morels appear in spring, European mushroom hunters take to the woods to collect *Hygrophorus marzuolus* near melting snowbanks. In North America, *Hygrophorus marzuolus* is known only from the Pacific Northwest.

Most edible waxcaps are members of the genus *Hygrophorus.* Many of them are fairly substantial, and none is known to be poisonous. In contrast, hygrocybes are typically too small and flavorless to be attractive as food. Exceptions include varieties of *Hygrocybe virginea* and *H. pratensis,* which are well-established edibles. *Hygrocybe punicea* is reported to be good, but we have not tried it. *Hygrocybe conica* and *H. flavescens* are reported to be poisonous. The edibility of many others in the genus is not known.

Although waxcaps are generally considered to be a safe group, we recommend consuming only those that have a tradition of culinary use. The edibility of many (especially in the genus *Hygrocybe*) remains unknown. As is the case with other wild mushrooms, some people may have an adverse reaction to species that are normally accepted as edible.

SIMILAR SPECIES THAT MAY BE MISTAKEN FOR WAXCAPS

Several mushrooms superficially resemble waxcaps. For example, *Lactarius hygrophoroides* is similar in stature to *Hygrophorus* but has nonwaxy gills, ornamented spores, and white latex that exudes from broken tissue. Some species of *Tricholoma* and *Clitocybe* are also similar in stature to *Hygrophorus,* but they have thin, nonwaxy gills. Species of *Laccaria* have waxy gills, but they are not sharp-edged, and their spores are distinctly spiny. The entire fruiting body of *Russula earlii* is very waxy in appearance, but it has brittle flesh and ornamented, amyloid spores. Some small, brightly colored chanterelles, such as *Cantharellus minor, C. ignicolor,* and *C. cinnabarinus,* are reminiscent of *Hygrocybe,* but their fertile surface is composed of blunt-edged folds or ridges rather than of true gills. *Mycena epipterygia* might be mistaken for a *Hygrocybe,* but it does not have waxy gills. The same can be said of certain species of *Nolanea,* such as *Nolanea quadrata* or *N. murraii,* which have thin, nonwaxy gills and angular, pink spores. Perhaps the most waxcap-like of all other mushrooms are members of the genera *Gomphidius* and *Chroogomphus.* However, these mushrooms can easily be distinguished by their dark gray to blackish spores.

Species Descriptions

HYGROCYBE

Hygrocybe acuta var. ***americana*** (A. H. Smith and Hesler) comb. nov.

Not Illustrated

SYNONYMS: *Hygrophorus acutus* A. H. Smith and Hesler

COMMON NAME: none.

CAP: 3–4.5 cm wide, broadly convex with a sharp conic umbo and a somewhat flaring margin; surface subviscid to viscid, glabrous, cuticle separable in shreds, short striate, dull lead gray over the umbo, pale brownish toward the margin; flesh waxy, brittle, whitish to grayish; odor and taste not distinctive.

GILLS: adnate to adnate-decurrent, close to subdistant, seceding at times, intervenose, whitish to pale ashy gray, not staining when handled or bruised; edges even.

STALK: 7–9 cm long, 6–8 mm thick, nearly equal overall, dry, solid, glabrous or slightly scurfy near the apex, white overall, not staining when handled or bruised.

MICROSCOPIC FEATURES: spores 6–8 × 4–5.5 µm, ellipsoid, smooth, hyaline, inamyloid.

OCCURRENCE: solitary on the ground under conifers; September–October; reported from Michigan, Oregon, and California; uncommon.

EDIBILITY: unknown.

OBSERVATIONS: The combination of a sharply conic, lead gray cap and white stalk is distinctive. *Hygrocybe acuta* F. H. Møller is a similar species with a brilliant red cap collected in the Faroe Islands. Compare with *Hygrocybe fornicata* (p. 29). The name *acuta* means "acutely pointed."

Hygrocybe acutoconica var. ***acutoconica*** (Clements) Singer Illus. p. 118

SYNONYMS:

Hygrocybe persistens var. *persistens* (Britzelmayr) Singer

Hygrophorus acutoconicus var. *acutoconicus* (Clements) A. H. Smith

COMMON NAME: none.

CAP: 2–10 cm wide, obtusely to sharply conic when young, soon campanulate; surface viscid to glutinous, glabrous, orange-red, orange, ochraceous orange, or yellow; margin often upturned, wavy, and splitting radially at maturity; flesh soft, yellow; odor and taste not distinctive.

GILLS: narrowly adnexed at first, becoming free at maturity, moderately broad, close to subdistant, yellow; edges entire or becoming serrated.

STALK: 6–12 cm long, 3–12 mm thick, nearly equal or enlarged toward the base, rounded or compressed, slippery, solid or hollow, fibrillose or glabrous, striate, sometimes twisted-striate, often splitting lengthwise, colored like the cap or paler, typically with a white base, not blackening when bruised, but often blackening at the base in age.

MICROSCOPIC FEATURES: spores 9–15 × 5–9 µm, ellipsoid, smooth, hyaline, inamyloid.

OCCURRENCE: solitary, scattered, or in groups on soil or among grasses in fields, along roadsides, or in broadleaf and mixed woodlands; April–January; widely distributed in the Southeast; occasional.

EDIBILITY: reported to be edible.

OBSERVATIONS: *Hygrocybe acutoconica* var. *cuspidata* (p. 12) is very similar, but it has a shiny, brilliant red cap. Also compare with *Hygrocybe conica* var. *conica* (p. 26), which is nearly identical but all parts stain black when bruised. *Hygrocybe acutoconica* var. *microspora* (Hesler and A. H. Smith) S. A. Cantrell and Lodge differs from *H.* var. *acutoconica* by having smaller spores that measure 7–10 × 5–6 µm. *Hygrocybe ruber* (Peck) comb. nov. is also similar, but according to Peck it has a red stalk that matches the cap color, smaller spores that measure 7–9 × 4.5–6 µm, and differently colored gills. The name *acutoconica* means "sharply cone-shaped."

Hygrocybe acutoconica var. ***cuspidata*** (Peck) Arnolds — Illus. p. 119

SYNONYMS:

Hygrocybe persistens var. *persistens* (Britzelmayr) Singer
Hygrophorus cuspidatus Peck
Hygrocybe cuspidata (Peck) Hongo and Izawa

COMMON NAME: none.

CAP: 2–7 cm wide; sharply conic when young, becoming broadly conic to nearly flat with an umbo in age; surface smooth, viscid when moist, shiny when dry, brilliant red to scarlet overall, fading to orange-red or paler in age; margin often uplifted and split in age; flesh thin, fragile, pale yellow to nearly white; odor and taste not distinctive.

GILLS: adnexed to free from the stalk, close to crowded, waxy, orange to yellow-orange to yellow.

STALK: 5–9 cm long, 5–10 mm thick, equal or enlarged slightly downward, stuffed to hollow, smooth, not viscid, often longitudinally striate or twisted-striate; orange-yellow, reddish yellow, or yellow, whitish at the base, not blackening in age or when bruised.

MICROSCOPIC FEATURES: spores 8–12 × 4–6.5 µm, ellipsoid, smooth, hyaline, inamyloid.

OCCURRENCE: in groups on the ground in mixed woods, roadsides, and burn sites; June–December; widely distributed; occasional.

EDIBILITY: unknown.

OBSERVATIONS: The brilliant red color of the shiny viscid cap and lack of staining when bruised are distinctive features. *Hygrocybe acutoconica* var. *acutoconica* (p. 11) is similar, but its cap color is orange-red to orange or yellow. *Hygrocybe ruber* (Peck) comb. nov. is also similar, but according to Peck it has a red stalk that matches the cap color, smaller spores that measure 7–9 × 4.5–6 µm and differently colored gills. The name *cuspidata* means "pointed."

Hygrocybe andersonii Cibula and N. S. Weber Illus. p. 122

SYNONYMS: none.

COMMON NAME: Clustered Dune Hygrocybe.

CAP: 1.3–3.3 cm wide, convex with a flattened to depressed center; surface smooth or often with a scurfy disc; color variable from orange-yellow, orange, or deep reddish orange to scarlet, becoming reddish brown to nearly black in age, disc often blackish; margin incurved; flesh thin, orange-yellow to reddish orange; odor and taste not distinctive.

GILLS: adnate to sinuate or subdecurrent, subdistant, fairly broad, yellow-orange to brownish orange or deep orange, becoming blackish in age.

STALK: 2.5–4.2 cm long, 6–9 mm thick, cylindrical to flattened, glabrous, smooth, often with longitudinal furrows, solid or hollow; colored like the cap or paler on the upper portion, becoming yellowish orange to pale yellow at the base.

MICROSCOPIC FEATURES: spores 16–19 × 3.8–5.6 µm, rod-shaped with a distinct projection (apiculus), smooth, hyaline, inamyloid.

OCCURRENCE: occasionally single but usually gregarious in caespitose clusters in sand near Florida Rosemary *(Ceratiola ericoides);* November–March; coastal areas and barrier islands along the Gulf Coast from Florida to Mississippi; locally common.

EDIBILITY: unknown.

OBSERVATIONS: This brightly colored waxcap is typically found in old, tree- and shrub-colonized sand dunes. It is distinctive for its clustered manner of growth and because it invariably occurs within the rhizosphere of the evergreen shrub Florida Rosemary. The species epithet honors the artist/naturalist Walter Inglis Anderson (1903–1965), who depicted this interesting waxcap on Horn Island, Mississippi, thirty-five years before it was scientifically described.

Hygrocybe angustifolia (Murrill) Candusso Illus. p. 83

SYNONYMS:

Camarophyllus angustifolius Murrill

Hygrophorus angustifolius (Murrill) Hesler and A. H. Smith

COMMON NAME: none.

CAP: 2–5 cm wide, obtuse to nearly plane, sometimes broadly umbonate, often slightly depressed on the disc with an uplifted margin at maturity; surface dry, unpolished, white; flesh thick and firm, white; odor and taste not distinctive.

GILLS: decurrent, close to subdistant, narrow, fairly thick, very brittle, sometimes forked; white; edges even.

STALK: 2–4 cm long, 1–2 cm wide, nearly equal or slightly tapered downward, solid, dry, glabrous or finely fibrillose, white.

MICROSCOPIC FEATURES: spores 3.5–6 × 3–4.5 µm, lacrymoid to subglobose or ellipsoid, smooth, hyaline, inamyloid.

OCCURRENCE: solitary, scattered, or in groups on the ground in mixed woodlands; August–December; widely distributed in the Northeast; uncommon.

EDIBILITY: unknown.

OBSERVATIONS: *Hygrocybe pratensis* var. *pallida* (Cooke) Arnolds (p. 85) is similar but has a wider cap and larger spores. *Hygrocybe virginea* var. *virginea* (p. 54) is also similar, but its cap is moist to lubricous, and the cap margin is somewhat striate and often remains decurved well into maturity. The name *angustifolia* means "narrow-leaved," referring to the narrow gills.

Hygrocybe appalachianensis (Hesler and A. H. Smith) comb. nov.

Illus. p. 119

SYNONYMS: *Hygrophorus appalachianensis* Hesler and A. H. Smith

COMMON NAME: Appalachian Waxcap.

CAP: 2–7 cm wide, convex, becoming depressed to funnel-shaped; surface fibrillose to fibrillose-scaly, dry or moist and slippery but not viscid; bright red, fading to reddish orange to orange-yellow; margin incurved when young, becoming uplifted and wavy in age; flesh yellowish, tinged orange; odor and taste not distinctive.

GILLS: adnate at first, becoming decurrent, subdistant to distant; whitish at first, soon becoming pale orange-yellow, often with a yellowish margin.

STALK: 3–7 cm long, 4–12 mm thick, nearly equal or tapered at the base, smooth or slightly scaly, moist or dry, not viscid, hollow, red on the upper portion, paler below.

MICROSCOPIC FEATURES: spores 11–18 × 7–10 µm, ellipsoid, smooth, hyaline, inamyloid; cheilocystidia and pleurocystidia 32–58 × 7–14 µm, clavate to spathulate.

OCCURRENCE: solitary or more often scattered in groups or caespitose clusters on the ground in mixed woods; June–December; widely distributed; occasional to locally common.

EDIBILITY: reported to be edible.

OBSERVATIONS: This waxcap's size and very large spores and the presence of cystidia are features that differentiate it from *Hygrocybe miniata* var. *miniata* (p. 36), *H. coccinea* (p. 23), *H. cantharellus* f. *cantharellus* (p. 20), and *H. coccineocrenata* (p. 23). The name *appalachianensis* refers to the Appalachian Mountains, the place where this waxcap was first described.

Hygrocybe atro-olivacea (A. H. Smith and Hesler) comb. nov. Not Illustrated

SYNONYMS: *Hygrophorus atro-olivaceus* A. H. Smith and Hesler

COMMON NAME: none.

CAP: 1–3 cm wide, convex to flat, center depressed at times, margin sometimes decurved and crenate; surface blackish brown at the center, olive brown at the margin, fading to gray-brown, moist, hygrophanous, squamulose, more so near the margin, translucent-striate to opaque when moist; flesh watery, brittle, soft; odor and taste not distinctive.

GILLS: adnate to subdecurrent, concolorous with cap margin, whitish when old, distant to subdistant, broad; edges may bruise olivaceous.

STALK: 3–5 cm long, 5–12 mm thick, gray brown, paler than the cap, glabrous, moist, often compressed.

MICROSCOPIC FEATURES: spores 5–6 × 3.5–4.5 µm, ellipsoid to subglobose, smooth, hyaline, inamyloid; cap and gill trama with granules within the cells that stain dark amber brown in Melzer's reagent.

OCCURRENCE: on the ground, often in caespitose clusters in woods, July–September; Michigan and Washington, distribution limits yet to be determined; uncommon.

EDIBILITY: unknown.

OBSERVATIONS: The small spores, gill trama reaction in Melzer's reagent, and dark brown coloration are distinctive features of this uncommon waxcap. The name *atro-olivacea* refers to its blackish and olive cap colors.

Hygrocybe aurantiosplendens R. Haller Illus. p. 103

SYNONYMS: *Hygrophorus aurantiosplendens* (R. Haller) P. D. Orton

COMMON NAME: none.

CAP: 1.5–5 cm wide, broadly conical to nearly plane with a broad umbo; margin entire or sometimes undulating or split; surface distinctly lubricous to viscid or somewhat glutinous, scarlet to orange-red or orange-yellow becoming more yellowish with age, sometimes with a broad reddish margin, hygrophanous, short translucent-striate from the margin; flesh colored like the cap surface or whitish, fragile; odor and taste not distinctive.

GILLS: narrowly adnate, sometimes emarginate with a decurrent tooth, subdistant, yellow to orange-yellow but typically paler than the cap.

STALK: 3–9 cm long, 5–18 mm thick, slightly enlarged downward or nearly equal, often distinctly tapered near the base, occasionally compressed, hollow at maturity, glabrous overall or sometimes whitish pruinose near the apex, moist, lubricous to somewhat viscid, typically paler yellow than the cap or concolorous, sometimes white at the base.

MICROSCOPIC FEATURES: spores 7.5–10 × 4–6 µm, ellipsoid, oblong or phaseoliform, frequently constricted near the middle, smooth, hyaline, inamyloid.

OCCURRENCE: scattered or in groups on the ground in mixed broadleaf woodlands; August–October; New York, distribution limits and frequency yet to be determined.

EDIBILITY: unknown.

OBSERVATIONS: Compare with *Hygrocybe flavescens* (p. 28), which has a less viscid stalk and lacks constricted spores but is otherwise very similar. The name *aurantiosplendens* means "gleaming orange."

Hygrocybe auratocephala (Ellis) comb. nov. Illus. p. 106

SYNONYMS:

Camarophyllus auratocephalus (Ellis) Saccardo
Hygrophorus auratocephalus (Ellis) Murrill

COMMON NAME: none.

CAP: 3–5 cm, conic, then campanulate with distinct umbo, splitting radially at the margin; surface glabrous, viscid when moist, bright yellow at first, then fading to pale yellow; flesh thin, yellow, becoming white in age; odor slightly to strongly mephitic; taste unpleasant.

GILLS: adnate with a decurrent tooth, very broad, subdistant to distant, ventricose, yellow.

STALK: 7–10 cm long, 2–7 mm thick, tapered downward, hollow, fragile, yellow, at times whitish at the base.

MICROSCOPIC FEATURES: spores 7–9 × 4–6 µm, ellipsoid to subovoid, smooth, hyaline, inamyloid.

OCCURRENCE: scattered on soil or among mosses in conifer or mixed woods; July–September; widely distributed in the Northeast; uncommon.

EDIBILITY: unknown.

OBSERVATIONS: Compare with *Hygrocybe mephitica* (p. 35), which has grayish violaceous to grayish purple gills and larger spores that measure 8.5–12 × 5–7 µm. It also has a skunk-like odor. Compare also with *Hygrocybe marginata* var. *concolor* (p. 33), which lacks a distinctive odor. The name *auratocephala* means "orangish head," referring to the cap color.

Hygrocybe basidiosa (Peck) comb. nov. Not Illustrated

SYNONYMS:

Camarophyllus basidiosus (Peck) Murrill

Hygrophorus basidiosus Peck

COMMON NAME: none.

CAP: 1–4 cm wide, convex to nearly plane, sometimes subumbonate; surface glabrous or appearing glaucous, hygrophanous, grayish brown when moist, fading to pale gray and becoming radiate-streaked; flesh whitish; odor and taste not distinctive.

GILLS: adnate to short decurrent, subdistant, broad, arched, thick, pale gray to dull gray; edges even.

STALK: 2.5–5 cm long, 3–10 mm thick, tapered downward or nearly equal, solid when young and becoming hollow in age, glabrous, moist or dry, white.

MICROSCOPIC FEATURES: spores 4–6 × 3–4.5 µm, subglobose, smooth, hyaline, inamyloid.

OCCURRENCE: scattered in woods, swamps, and sphagnum bogs; July–October; widespread in the Northeast; uncommon.

EDIBILITY: unknown.

OBSERVATIONS: The combination of a grayish brown cap that fades to pale gray, pale gray to dull gray gills, and white stalk is distinctive. *Hygrocybe albipes* (Peck) comb. nov., reported from Maine, Massachusetts, and Alabama, is similar but has a smaller grayish brown cap with a strongly decurved margin, strongly decurrent whitish gills that darken in age, a solid white stalk that is attenuated at the base, and larger subglobose to broadly ellipsoid or ovoid spores that measure 5.5–8 × 4.5–6 µm. The species name is derived from the Latin word *basidi,* which means "small pedestal."

Hygrocybe caespitosa Murrill Illus. p. 87

SYNONYMS:

Camarophyllus caespitosus Murrill

Hygrophorus caespitosus (Murrill) Murrill

COMMON NAME: Clustered Waxcap.

CAP: 1–6 cm wide, broadly convex to nearly flat, sometimes depressed on the disc; surface moist at first, soon dry, not viscid, covered with small brownish olive scales with darker brown to blackish tips over a whitish to yellowish or honey yellow ground color; margin often uplifted and torn in age; flesh thick, yellowish; odor not distinctive; taste resembling raw potato or not distinctive.

GILLS: adnate to slightly decurrent, subdistant to distant, broad, sometimes forked, waxy, whitish to pale yellow.

STALK: 2–5 cm long, 3–7 mm thick, nearly equal or tapered downward slightly, smooth, dry, becoming hollow, pale yellow with olive yellow to yellow tones near the base.

MICROSCOPIC FEATURES: spores 6.5–10 × 4–7 µm, ellipsoid, smooth, hyaline, inamyloid.

OCCURRENCE: scattered or in groups, often in caespitose clusters on the ground under conifers and broadleaf trees; June–December; widely distributed; occasional to fairly common locally.

EDIBILITY: unknown.

OBSERVATIONS: The presence of small dark scales over whitish to honey yellow caps and the growth in clusters make this waxcap easy to recognize in the field. The epithet *caespitosus* refers to the clustered growth pattern, which is typical for this species.

Hygrocybe calciphila Arnolds — Not Illustrated

SYNONYMS: *Pseudohygrocybe calciphila* (Arnolds) Kovalenko

COMMON NAME: none.

CAP: 4–25 mm wide, hemispherical to convex when young, becoming nearly plane and often with a slightly depressed center at maturity; surface moist or dry and finely squamulose, sometimes nearly pruinose over the disc, scarlet to orange-red, often with a narrow yellow margin, fading to dull orange, orange-yellow, or yellow in age; squamules colored like the cap but soon turning yellow or grayish after drying; margin sometimes indistinctly translucent-striate; flesh thin, colored like the cap or paler; odor and taste not distinctive.

GILLS: broadly adnate or rarely subdecurrent, close to subdistant, whitish to pale yellow when young, soon becoming yellowish to pale orange or pinkish orange; edges entire.

STALK: 2–7 cm long, 1.5–10 mm thick, nearly equal or narrowed downward near the base, cylindrical or compressed and often with a longitudinal groove, moist or dry but not viscid, orange or golden yellow or a mixture of these colors, sometimes white at the base.

MICROSCOPIC FEATURES: spores 6–11 × 4–7 µm, broadly ellipsoid, sometimes subglobose or ellipsoid, not constricted or widened near the base, smooth, hyaline, inamyloid, mostly binucleated.

OCCURRENCE: solitary, scattered, or in groups in grassy areas and woodlands typically associated with calcareous soils; July–October; New England and New York, distribution limits and frequency yet to be determined.

EDIBILITY: unknown.

OBSERVATIONS: *Hygrocybe miniata* var. *miniata* (p. 36) is nearly identical but has ellipsoid, uninucleated spores that are often constricted in the middle and widened near the base. Compare with *Hygrocybe squamulosa* (p. 51), which has minute to distinct squamules especially near the margin and smaller spores that are uninucleated. The name *calciphila* means "limestone loving."

Hygrocybe calyptriformis (Berkeley) Fayod Illus. p. 124

SYNONYMS:

Hygrophorus calyptriformis Berk. and Broome in Berkeley
Hygrophorus amoenus (Lasch) Quélet

COMMON NAME: Pink Waxcap.

CAP: 2.5–6 cm wide, sharply conic when young, becoming broadly convex with a sharply pointed umbo in age; surface smooth, slightly viscid when wet, coral red to brownish pink or vinaceous pink when young, fading to coral pink to shell pink or salmon pink to pinkish buff in age; margin often split and lobed in age; flesh thin, whitish pink; odor and taste not distinctive.

GILLS: adnexed to nearly free, close to subdistant, narrow, waxy, pinkish buff to cream with lavender tints.

STALK: 5.5–14 cm long, 3–7 mm thick, equal, hollow, smooth, moist or dry but not viscid, fragile, whitish overall when young, becoming lavender to pinkish on the upper two-thirds in age.

MICROSCOPIC FEATURES: spores 6–9 × 4–6 µm, ellipsoid, smooth, hyaline, inamyloid; pleurocystidia scattered, 50–140 × 8–25 µm, clavate.

OCCURRENCE: scattered or in groups on the ground, often in grassy areas under conifers and broadleaf trees; July–November; widely distributed; occasional.

EDIBILITY: reported to be edible.

OBSERVATIONS: The beautiful pink cap coloration and the very large clavate pleurocystidia are distinctive features. Compare with *Hygrocybe pura* (p. 47), which is white but may have pinkish or reddish staining. The name *calyptriformis* means "resembling a hood," in reference to the cap shape.

Hygrocybe canescens (A. H. Smith and Hesler) P. D. Orton Illus. p. 97

SYNONYMS:

Camarophyllus canescens (A. H. Smith and Hesler) Singer
Hygrophorus canescens A. H. Smith and Hesler

COMMON NAME: none.

CAP: 2–5 cm wide, obtuse, becoming broadly convex or nearly plane, sometimes with a low broad umbo; margin incurved and remaining so well into maturity, often lobed or irregular; surface dry and tomentose in patches and along the margin, bluish gray to pale gray with paler grayish white areas where the tomentose patches remain, fading somewhat in older specimens, mostly opaque with scattered spots as specimens age, not translucent-striate and only sparingly hygrophanous; flesh pale gray, thin, not staining when cut, fragile; odor slightly unpleasant or not distinctive; taste not distinctive.

GILLS: decurrent, subdistant to distant, narrow, sometimes forking, intervenose, light gray to bluish gray, fading as specimens dry.

STALK: 4–6 cm long, 6–10 mm thick, typically tapered downward, sometimes twisted, often silky shiny, dry and tomentose, hollow at maturity, pale bluish gray covered by a white tomentose layer.

MICROSCOPIC FEATURES: spores 4–6 × 4–4.5 µm, broadly ellipsoid to subglobose, smooth, hyaline, inamyloid.

OCCURRENCE: solitary, scattered, or in groups on the ground or among mosses usually under conifers, sometimes in grassy areas; August–December; widely distributed; uncommon.

EDIBILITY: unknown.

OBSERVATIONS: *Hygrocybe lacmus* (p. 32) is similar but lacks the tomentose layer on the cap and stalk, has a lubricous or viscid cap surface, and has somewhat larger spores. The name *canescens* means "hoary white," in reference to the cap's appearance.

Hygrocybe cantharellus* f. *cantharellus (Schweinitz) Murrill Illus. p. 109

SYNONYMS:

Camarophyllus cantharellus (Schweinitz) Murrill
Hygrocybe lepida Arnolds
Hygrophorus cantharellus (Schweinitz) Fries

COMMON NAME: Chanterelle Waxcap.

CAP: 1–3.5 cm wide; convex, becoming flat and finally depressed; surface dry, tomentose to finely scurfy at first, becoming finely squamulose to lacerate-squamulose especially on the disc, orange, scarlet, or orange-red, becoming paler in age; squamules appressed or sometimes erect, concolorous with the surface or paler, sometimes yellow as specimens dry; margin often recurved and undulating or crenate at maturity; flesh thin, colored like the cap or yellowish; odor sometimes resembling raw potatoes or not distinctive; taste not distinctive.

GILLS: typically strongly decurrent but sometimes only slightly so, subdistant to distant, broad, waxy, ivory to creamy yellow or orange-yellow; edges sometimes paler than the faces.

STALK: 3–9 cm long, sometimes longer when growing among sphagnum mosses, 2–5 mm thick, equal or enlarged slightly upward, stuffed or hollow, smooth, dry, fragile, colored like the cap or paler, often whitish or pale yellow at the base.

MICROSCOPIC FEATURES: spores 7–12 × 4–8 µm, ellipsoid to nearly oval, uninucleate, smooth, hyaline, inamyloid.

OCCURRENCE: occasionally solitary but usually in groups or clustered on rich humus, on decaying wood or among mosses, often in bogs; July–January; widely distributed; fairly common.

EDIBILITY: reported to be edible but of little substance.

OBSERVATIONS: The tall slender stature of this species is distinctive. *Hygrocybe coccineocrenata* (p. 23) is quite similar but has squamules on the cap that turn grayish brown to dark brown with age. Also compare with *Hygrocybe squamulosa* (p. 51), which has conspicuous appressed squamules on the cap except for the margin, broadly adnate gills, and predominantly binucleate spores. The species name refers to the similarity in form to chanterelle mushrooms (*Cantharellus* species).

Hygrocybe ceracea (Fries : Fries) P. Kummer — Not Illustrated

SYNONYMS:

Hygrocybe subceracea Murrill
Hygrocybe nitida var. *lutea* Murrill
Hygrophorus ceraceus (Fries) Fries

COMMON NAME: none.

CAP: 1–5 cm wide, convex to hemispherical, becoming broadly convex or obtuse, the disc sometimes flattened or slightly depressed; margin sometimes upturned and wavy or crenulate; surface lubricous, glabrous, hygrophanous, translucent-striate nearly to the disc at maturity when moist, often with a conspicuous translucent eyespot on the disc of young specimens when moist, bright yellow to orange-yellow; flesh thin, very soft and fragile, yellowish; odor and taste not distinctive.

GILLS: broadly adnate to slightly decurrent, subdistant, typically pale yellow but sometimes whitish or darker orange-yellow; edges even.

STALK: 2–5 cm long, 1.5–4 mm thick, nearly equal, sometimes compressed, glabrous, slightly viscid when moist but soon dry, hollow, colored like the cap.

MICROSCOPIC FEATURES: spores 5.5–8 × 3–5.5 µm, broadly ovoid to subellipsoid, sometimes constricted near the midsection, smooth, hyaline, inamyloid.

OCCURRENCE: scattered or in groups in grassy areas or on soil or among mosses in broadleaf or mixed woods; July–January; widely distributed; occasional.

EDIBILITY: unknown.

OBSERVATIONS: *Hygrocybe ceracea* is one of several confusing small, more or less viscid to lubricous yellow hygrocybes. Compare with *Hygrocybe nitida* (p. 40), which has decurrent gills and a distinctly viscid, conspicuously depressed to infundibuliform cap that usually fades to pale yellow or whitish at maturity. *Hygrocybe chlorophana* (p. 22) is also similar, but it has adnexed gills, a viscid cap, and a viscid stalk. The name *ceracea* means "waxy."

Hygrocybe chamaeleon (Cibula) comb. nov. Illus. p. 118

SYNONYMS: *Hygrophorus chamaeleon* Cibula

COMMON NAME: Chamaeleon Waxcap.

CAP: 1–4 cm wide, convex to flat, becoming depressed in the center, at times funnel-shaped, moist, not viscid, rimose at times, center fibrillose-squamulose; surface variable in color, exposed caps dark red to reddish brown at the margin when moist, drying to reddish brown to brownish orange, litter-covered caps pale yellow to olive green or grayish; flesh yellowish to reddish, thin, hygrophanous.

GILLS: adnate to decurrent, subdistant to distant, variable in color, pale yellow, grayish, reddish orange, greenish yellow; edges serratulate.

STALK: 3–6.5 cm long, 5–9 mm thick, terete, hollow, dry or moist but not viscid, glabrous, with a shiny luster red, apex dark red to reddish orange, becoming yellowish below, base light yellow to cream.

MICROSCOPIC FEATURES: as recorded in the original description, spores highly variable in size, some 12–17 × 7–10 µm and others 6.5–8.5 × 4–5.5 µm, ellipsoid to oval, smooth, hyaline, inamyloid.

OCCURRENCE: gregarious to clustered in leaf litter in low areas under magnolia and in mixed woods; July–September; Mississippi and Texas, distribution limits yet to be determined; locally abundant.

EDIBILITY: unknown.

OBSERVATIONS: The spores of this variably colored waxcap are usually on the larger end of the size range. It is likely to be encountered throughout the Gulf Coast region. The species epithet refers to the variable cap color.

Hygrocybe chlorophana (Fries : Fries) Wünsche Illus. p. 100

SYNONYMS: *Hygrophorus chlorophanus* (Fries) Fries

COMMON NAME: Golden Waxcap.

CAP: 2–7 cm wide, hemispherical, becoming broadly convex to nearly plane in age, sometimes with a low broad umbo or depressed on the disc; surface viscid, glabrous, lemon yellow to orange-yellow; margin weakly translucent-striate; flesh thin, pale yellow to yellow; odor and taste not distinctive.

GILLS: adnexed, close, broad, whitish at first, becoming yellowish to yellow-orange at maturity; edges even.

STALK: 3–8 cm long, 4–12 mm thick, nearly equal, often compressed with one or two longitudinal grooves, distinctly viscid when fresh but sometimes dry and shiny, hollow at maturity, colored like the cap or slightly paler.

MICROSCOPIC FEATURES: spores 6–10 × 4–6.5 µm, ellipsoid to ovoid, or oblong, smooth, hyaline, inamyloid.

OCCURRENCE: solitary, scattered, or in groups on the ground in woodlands; June–December; widely distributed; occasional.

EDIBILITY: reported to be edible.

OBSERVATIONS: *Hygrocybe flavescens* (p. 28) is very similar, but it has an orange to yellow-orange cap and a lubricous to moist or dry stalk that is not distinctly viscid. Compare with *Hygrocybe ceracea* (p. 21). The name *chlorophana* means "appearing greenish," which seems a bit of a stretch, at least in our experience.

Hygrocybe coccinea (Schaeffer : Fries) P. Kummer Illus. p. 120

SYNONYMS: *Hygrophorus coccineus* (Schaeffer : Fries) Fries

COMMON NAME: Scarlet Waxcap.

CAP: 2–7 cm wide, conic when young, becoming convex to flat with an umbo in age; surface moist or dry, smooth, not viscid, not fibrillose-scurfy when dry, bright red when young, fading to orange-red in age; flesh thin, fragile, red to reddish orange, slowly staining gray in $FeSO_4$; odor and taste not distinctive.

GILLS: adnate, close to subdistant, broad, waxy, yellowish, red-orange to creamy peach, often with a yellowish margin.

STALK: 2–8 cm long, 2–10 mm thick, equal, hollow, frequently compressed, dry, smooth, yellow-orange to orange-red, paler yellow toward the base, and often coated with white mycelium at the base.

MICROSCOPIC FEATURES: spores 7–11 × 4–5 µm, ellipsoid, smooth, hyaline, inamyloid.

OCCURRENCE: solitary, scattered, or in groups on the ground in conifer and broadleaf forests; July–November; widely distributed; fairly common.

EDIBILITY: edible by most accounts.

OBSERVATIONS: *Hygrocybe punicea* (p. 46) is similar but typically has a larger cap that is viscid and tends to be deeper blood red, especially when young; it also has pale yellow to whitish flesh, narrowly adnate gills that are sometimes emarginate with a decurrent tooth, and a coarsely fibrillose, striate stalk. Much confusion exists about the similar *Hygrocybe marchii* (Bresadola) Singer, which has been variously described in Europe and North America. Some authors describe the cap as greasy to lubricous, whereas others state that it is viscid then soon dry. Numerous additional differences, both macroscopic and microscopic, exist among investigators' descriptions. Several authors suggest that *Hygrocybe marchii* can be reduced to varietal status. We are not convinced that *Hygrocybe marchii,* as described by European mycologists, exists in North America and therefore have chosen not to include it in this work. The species name comes from the Latin term *coccin,* which means "scarlet."

Hygrocybe coccineocrenata (P. D. Orton) M. M. Moser Illus. p. 110

SYNONYMS:

Hygrocybe coccineocrenata var. *sphagnophila* (Peck) Arnolds

Hygrophorus coccineocrenatus P. D. Orton
Hygrophorus miniatus var. *sphagnophilus* Peck
Hygrophorus turundus var. *sphagnophilus* (Peck) Hesler and A. H. Smith

COMMON NAME: Sphagnum Waxcap.

CAP: 6–20 mm wide, convex when young, becoming nearly plane at maturity, often with a depressed center; margin decurved well into maturity, often crenate and translucent-striate when moist; surface moist at first then soon dry, hygrophanous, minutely squamulose especially over the disc, color variable from bright red to orange-red, yellow-orange, or yellow; squamules at first concolorous with the surface, sometimes remaining concolorous but typically turning pale grayish brown to dark brown, appressed or erect; flesh thin, colored like the cap surface or paler; odor and taste not distinctive.

GILLS: short-decurrent to decurrent, subdistant to distant, narrow, fairly thick, whitish or pale yellow; edges even.

STALK: 1.5–7 cm long, 1.5–5 mm thick, tapered downward or nearly equal, sometimes conspicuously elongated when growing among sphagnum mosses, moist or dry, red to orange-red and often paler or yellow down toward the base, fading in age, sometimes with orange hairs at the base.

MICROSCOPIC FEATURES: spores 8–12 × 5.5–8 µm, ellipsoid to oblong or obovoid-oblong, smooth, hyaline, inamyloid.

OCCURRENCE: solitary, scattered, or in groups in moist habitats, often among sphagnum mosses; widespread in the Northeast, distribution limits and frequency of occurrence yet to be determined.

EDIBILITY: unknown.

OBSERVATIONS: This small waxcap is often confused with the European *Hygrocybe turunda* var. *turunda* (Fries) P. Kaesten, which may or may not occur in North America. *Hygrocybe cantharellus* f. *cantharellus* (p. 20) is similar, but its cap squamules are concolorous with the cap surface, not grayish brown to dark brown at maturity. The species name is a combination meaning "scarlet" and "notched," referring to the cap color and the crenulate cap margin.

Hygrocybe colemanniana (A. Bloxam) P. D. Orton and Watling Illus. p. 93

SYNONYMS:

Camarophyllus colemannianus (A. Bloxam) Ricken
Hygrophorus colemannianus A. Bloxam in Berkeley and Broome
Hygrophorus uliginosus Hesler and A. H. Smith

COMMON NAME: none.

CAP: 1–5 cm wide, hemispherical to obtuse at first, becoming broadly convex to nearly plane, often distinctly umbonate, typically translucent-striate when fresh; margin incurved at first, becoming uplifted and somewhat wavy or irregular in age; surface moist to slightly lubricous when fresh, hygrophanous, glabrous and sometimes shiny, reddish brown to orange-brown, darker cinnamon brown, or date brown, paler toward the margin; flesh thick under the disc, thinner near the margin, colored like the cap surface but paler, not staining when exposed; odor and taste not distinctive.

GILLS: distinctly decurrent, close to subdistant, narrow to moderately broad, strongly intervenose, often forking, pale brown or whitish with a pinkish brown tint; edges entire.

STALK: 2–8 cm long, 3–7 mm thick, nearly equal or narrowed toward the base, dry and smooth to slightly fibrillose, solid or narrowly hollow, white to pale brown.

MICROSCOPIC FEATURES: spores 6–9 × 4.5–7 µm, ellipsoid to broadly ellipsoid, smooth, hyaline, inamyloid.

OCCURRENCE: solitary, scattered, or in groups on the ground in oak and beech woods, August–November; widely distributed in the Northeast; uncommon.

EDIBILITY: unknown.

OBSERVATIONS: The combination of a reddish brown cap, white to pale brown stalk, and ellipsoid to broadly ellipsoid spores is distinctive. *Hygrophorus uliginosus* Hesler and A. H. Smith is reported to differ by lacking a viscid cap, having slightly smaller spores and a grayish cap tinged with cinnamon, and growing on muck in cedar swamps. We have placed *Hygrophorus uliginosus* in synonymy because the differentiating characteristics are slight.

Hygrocybe conica var. ***atrosanguinea*** (Grund and Harrison) comb. nov.

Illus. p. 121

SYNONYMS: *Hygrophorus conicus* var. *atrosanguineus* (Fries) Grund and Harrison

COMMON NAME: Blood-Red Witch's Hat.

CAP: 1.5–7 cm wide, obtusely conical at first becoming broadly conical, usually with a distinct umbo; margin at first entire then becoming deeply incised, lobed, or cracked; surface slightly viscid when moist, otherwise dry, appressed-fibrillose and often appearing radially streaked, dark strawberry red to violet-brown, staining black when handled or bruised, soon blackening after picking; flesh thin, colored like the cap surface; odor usually strongly aromatic; taste not distinctive.

GILLS: adnexed to nearly free, close, fairly broad, waxy, mustard yellow to grayish orange when young, becoming olivaceous ocher at maturity, soon blackening after picking, handling, or bruising.

STALK: 4.5–10 cm long, 3–10 mm thick, nearly equal overall, longitudinally twisted-striate, fragile and sometimes splitting, minutely fibrillose, hollow, dry to moist, not viscid, reddish orange to mustard yellow down to a whitish base, bruising black.

MICROSCOPIC FEATURES: spores 8–10 × 5–6.5 µm, subellipsoid to ellipsoid, smooth, hyaline, inamyloid.

OCCURRENCE: solitary or scattered on the ground or in needle duff in hemlock woods; June–October; widespread in the Northeast; rare.

EDIBILITY: unknown.

OBSERVATIONS: *Hygrocybe conica* var. *conica* (p. 26) is similar but has a lighter-colored cap that lacks the radial streaks, and its flesh lacks an aromatic odor. Compare with *Hygrocybe conicoides* (p. 27), which is restricted to sandy coastal environments. The name *conica* means "cone-shaped," and the variety name *atrosanguinea* means "dark blood red."

Hygrocybe conica var. ***conica*** (Schaeffer: Fries) P. Kummer Illus. p. 121

SYNONYMS: *Hygrophorus conicus* (Fries) Fries

COMMON NAME: Witch's Hat.

CAP: 2–9 cm wide, sharply conic to campanulate, usually with an umbo; surface smooth, slightly viscid when moist, otherwise dry, dark orange-red to red, orange, lighter orange near the margin or sometimes yellow overall, often with olive green tints, slowly staining black when bruised or in age; flesh thin, fragile, colored like the cap, bruising black; odor and taste not distinctive.

GILLS: free from the stalk, close, broad, waxy, light yellow to greenish orange, becoming black in age or where bruised.

STALK: 2–10 cm long, 3–10 mm thick, equal, hollow at maturity, fragile, smooth, moist or dry but not viscid, often longitudinally striate or twisted-striate, yellow to yellow-orange, pale yellow near the base, staining black when bruised or in age.

MICROSCOPIC FEATURES: spores 8–10 × 5–5.6 µm, ellipsoid, smooth, hyaline, inamyloid.

OCCURRENCE: solitary to scattered on the ground in broadleaf woods, in open areas, and under conifers; June–January; widely distributed; occasional to fairly common.

EDIBILITY: not recommended, often reported as poisonous.

OBSERVATIONS: *Hygrocybe conicoides* (p. 27) differs in soon developing salmon orange to reddish gills and larger, more elongated spores and in having a sand-dune habitat. *Hygrocybe acutoconica* var. *acutoconica* (p. 11) has a sharply conic, orange-red, orange, ochraceous orange, or yellow cap and a similarly colored stalk that is longitudinally twisted or striate, often blackening at the base in age, but not blackening when bruised. *Hygrocybe coccinea* (p. 23) has a dry, bright red cap that fades to orange-red in age, red-orange to creamy peach or yellowish gills, and a yellow-orange to orange-red stalk that is paler yellow down toward the base and is often coated with white mycelium on the base. Compare also with *Hygrocybe singeri* var. *singeri*

(p. 50), which is similar but has a distinctly viscid stalk. The name *conica* refers to the cone-shaped cap.

Hygrocybe conicoides (Orton) P. D. Orton and Watling Illus. p. 123

SYNONYMS:

Hygrophorus conicoides P. D. Orton

Hygrophorus conicus var. *conicoides* (Orton) Arnolds

COMMON NAME: Dune Witch's Hat.

CAP: 2.5–6.5 cm wide, campanulate to convex or conical, becoming more or less flattened with an acute umbo in age; surface smooth, satiny when dry, scarlet to cherry red, slowly turning black in age; flesh thin, fragile, colored like the cap, slowly turning black on exposure; odor and taste not distinctive.

GILLS: adnexed, close, broad, waxy, yellow at first, soon infused with salmon orange to reddish tones, eventually becoming black.

STALK: 2.5–7.5 cm long, 3–10 mm thick, equal, longitudinally striate or twisted-striate, yellow to orange, often paler at the base, slowly turning black when handled or in age.

MICROSCOPIC FEATURES: spores 10–14 × 4–6 µm, elongated-ellipsoid, smooth, hyaline, inamyloid.

OCCURRENCE: solitary or in small groups in coastal sand dunes; September–January; widely distributed; occasional to fairly common locally.

EDIBILITY: unknown, possibly poisonous.

OBSERVATIONS: This attractive waxy cap is often buried "up to its neck" in sand. *Hygrocybe conica* var. *conica* (p. 26) is very similar, but its gills do not develop salmon orange to reddish tones, it has smaller spores, and it typically grows on soil in broadleaf or conifer woods or in open areas. The name *conicoides* means "cone-like."

Hygrocybe deceptiva (A. H. Smith and Hesler) Leelavathy, Manimohan and Arnolds Illus. p. 90

SYNONYMS:

Camarophyllopsis deceptivus (A. H. Smith and Hesler) Bon

Hygrophorus deceptivus A. H. Smith and Hesler

Armillaria deceptiva (A. H. Smith and Hesler) Singer

COMMON NAME: Deceptive Waxcap.

CAP: 1–4 cm wide, convex to flat with a slightly upturned margin; surface tawny brown to cinnamon brown, fuscous when wet; margin striate when wet; flesh pale or concolorous with the cap, thick at the center, thin toward the margin; odor not distinctive; taste slightly acidic or not distinctive.

GILLS: adnate to subdecurrent, close to subdistant, broad, intervenose, edges even, pale buff or concolorous with the cap.

STALK: 2–10 cm long, 3–10 mm thick, tapering downward, hollow, fragile, glabrous, concolorous with the cap; apex whitish, pruinose or canescent.

MICROSCOPIC FEATURES: spores 3–5 × 3–4 µm, ellipsoid to globose, smooth, hyaline, inamyloid.

OCCURRENCE: gregarious in humus and on soil in broadleaf and mixed woods; June–August; southeastern states, also known from Europe; infrequent.

EDIBILITY: unknown.

OBSERVATIONS: Compare with *Hygrocybe subovina* (p. 53) and *H. ovina* (p. 42), which are similar but have fragrant to sweet or nitrous odors, respectively. The name *deceptiva* means "deceptive," perhaps relating to the possible confusion in recognizing this rather nondescript waxcap.

Hygrocybe flavescens (Kauffman) Singer Illus. p. 101

SYNONYMS:

Hygrophorus flavescens (Kauffman) A. H. Smith and Hesler
Hygrophorus puniceus var. *flavescens* Kauffman

COMMON NAME: Golden Orange Waxcap.

CAP: 2.5–7 cm wide, broadly convex to nearly flattened or slightly depressed on the disc; surface viscid when fresh, becoming dry and shiny, glabrous, orange to yellow-orange, fading to paler yellow-orange to pale orange-yellow or amber yellow in age, rarely white throughout; margin incurved at first and often remaining so well into maturity, somewhat striate when moist; flesh thin, waxy, yellowish; odor and taste not distinctive.

GILLS: adnexed, broad, close to subdistant, yellow; edges even.

STALK: 4–7 cm long, 5–16 mm thick, nearly equal overall or narrowed near the base, often compressed and grooved on one side, fragile and easily split, glabrous or faintly fibrillose, lubricous to moist or dry but not viscid, hollow in age, colored like the gills near the apex, yellow-orange to orange downward to a whitish base.

MICROSCOPIC FEATURES: spores 7–9 × 4–5 µm, ellipsoid, smooth, hyaline, inamyloid.

OCCURRENCE: scattered or in groups on soil and humus in broadleaf, conifer, or mixed woods, frequently in grassy woodland margins; June–December; widely distributed; fairly common.

EDIBILITY: not recommended; variously reported to be either edible or poisonous.

OBSERVATIONS: The similar *Hygrocybe chlorophana* (p. 22) has a brighter lemon yellow cap and a distinctly viscid stalk. *Hygrocybe citrinovirens* (J. E. Lange) Jülich Schäffer, reported from grassy areas in Europe and possibly from broadleaf woodlands in North Carolina, may simply be a variety of

H. flavescens. It has a dry and narrowly conical orange-yellow to greenish yellow or lemon yellow cap with brownish fibers, a cap margin that is often lobed and splits radially, white to yellowish gills, and a dry, smooth to fibrillose stalk. The name *flavescens* means "turning yellow."

Hygrocybe fornicata (Fries) Singer Illus. p. 92

SYNONYMS:
Camarophyllus fornicatus (Fries) P. Karsten
Hygrophorus fornicatus Fries

COMMON NAME: none.

CAP: 2–8 cm wide, broadly conical at first, becoming nearly plane at maturity, frequently with an umbo or papilla when fully expanded; margin incurved when young, often uplifted, wavy, and split in older specimens, sometimes developing pale reddish stains when bruised; surface moist and somewhat lubricous or dry, smooth, glabrous or with tiny squamules especially over the center, variously colored, including brownish gray to grayish brown or pale brown or sometimes whitish, nearly always darker over the center and paler toward the margin; flesh firm, thin, waxy, white, not staining when exposed; odor and taste not distinctive.

GILLS: adnate to emarginate, close to crowded, thin, intervenose, white or tinted grayish; edges even or somewhat eroded, sometimes developing pale reddish stains when bruised.

STALK: 2–7 cm long, 6–20 mm thick, nearly equal overall or enlarged downward, moist or dry, fibrillose to finely squamulose especially near the base, hollow at maturity, white, sometimes developing pale reddish stains when bruised.

MICROSCOPIC FEATURES: spores 6–9.5 × 4.5–6 µm, broadly ellipsoid to ellipsoid or obovoid, smooth, hyaline, inamyloid.

OCCURRENCE: scattered or in groups on the ground in broadleaf woods; August–October; recorded from New York, distribution limits and frequency yet to be determined.

EDIBILITY: unknown.

OBSERVATIONS: We are unaware of any collections of this mushroom in North America other than in California and New York. The senior author accompanied world waxcap authority Dr. Eef Arnolds of the Netherlands when this mushroom was first collected in New York. The name *fornicata* means "arched," for the uplifted margin of mature caps.

Hygrocybe hondurensis (Murrill) comb. nov. Not Illustrated

SYNONYMS: *Hygrophorus hondurensis* (Murrill) Murrill

COMMON NAME: none.

CAP: 1–1.5 cm wide, convex to nearly plane, slightly depressed on the disc; surface red to yellowish, very viscid, radially striate; flesh thin, yellow; odor and taste unknown.

GILLS: subdecurrent, subdistant, narrow, whitish to yellowish.

STALK: 3–4 cm long, 1–2 mm thick, nearly equal, distinctly viscid, lemon yellow to deep yellow.

MICROSCOPIC FEATURES: spores 5–7 × 3–4.5 µm, ellipsoid, smooth, hyaline, inamyloid.

OCCURRENCE: scattered on soil; August–October; known only in eastern North America from Florida and Michigan; also reported from Washington, British Honduras, and Trinidad; possibly rare.

EDIBILITY: unknown.

OBSERVATIONS: Although reported from a wide range of locations, this species is not well known. The name *hondurensis* refers to the Central American country Honduras.

Hygrocybe huronensis (A. H. Smith and Hesler) Singer — Not Illustrated

SYNONYMS: *Hygrophorus huronensis* A. H. Smith and Hesler

COMMON NAME: none.

CAP: 1–3 cm wide, convex at first, becoming broadly convex to nearly plane in age; margin incurved and remaining so well into maturity; surface glabrous, hygrophanous, viscid, distinctly translucent-striate when fresh, watery white to snow white; flesh thin, firm, very waxy, watery white to snow white, not staining when bruised; odor and taste not distinctive.

GILLS: adnexed, subdistant, broad, thin, white, not staining when bruised; edges even.

STALK: 3–5 cm long, 3–5 mm thick, nearly equal overall, dry, glabrous, hollow at maturity, translucent and appearing glassy when moist, watery white to snow white.

MICROSCOPIC FEATURES: spores 7–9 × 4.5–6 µm, ellipsoid, smooth, hyaline, inamyloid.

OCCURRENCE: scattered on grassy soil under brush; September; currently reported only from Michigan, distribution and frequency yet to be determined.

EDIBILITY: unknown.

OBSERVATIONS: The combination of a viscid cap, dry stalk, adnexed gills, and white color throughout is distinctive. Singer (1951) reported that white forms observed among groups of *Hygrocybe flavescens* are indistinguishable from *H. huronensis*. The species name refers to the Huron River in Michigan, the area where the type specimen was first collected.

Hygrocybe hymenocephala (A. H. Smith and Hesler) P. D. Orton and Watling
Illus. p. 88

SYNONYMS:
Hygrophorus hymenocephalus A. H. Smith and Hesler
Camarophyllus hymenocephalus (A. H. Smith and Hesler) Lange
Camarophyllopsis hymenocephala (A. H. Smith and Hesler) Arnolds

COMMON NAME: none.

CAP: 5–30 mm wide, convex to orbicular, becoming plane; margin incurved and often crenate; surface glabrous, hygrophanous, moist or dry, pinkish cinnamon to pinkish brown, fading to clay brown or pinkish buff, becoming darker in age; flesh waxy, thick at the center, thin toward the margin, whitish or concolorous with the cap; odor and taste not distinctive.

GILLS: adnate to decurrent, distant, brown, often concolorous with the cap; edges even, often paler than the sides.

STALK: 2–8 cm long, 2–7 mm thick, nearly equal or tapered downward, solid, becoming hollow, fragile, glabrous below, canescent above, moist or dry, concolorous with the cap, becoming darker in age, at times nearly black.

MICROSCOPIC FEATURES: spores 4–6 × 4–5 µm, globose to subglobose, at times appearing ellipsoid, smooth, hyaline, inamyloid.

OCCURRENCE: scattered to gregarious on soil and decaying logs in mixed woods and under rhododendron; August–October; Nova Scotia to Michigan and south to North Carolina, also reported in Europe.

EDIBILITY: unknown.

OBSERVATIONS: *Hygrophorus peckianus* Howe is similar but has whitish to gray gills and an odor described as offensive. The name *hymenocephalus* means "having a membranous head."

Hygrocybe irrigata (Persoon : Fries) Bon Illus. p. 94

SYNONYMS:
Hygrocybe unguinosa (Fries : Fries) P. Karsten
Hygrophorus irrigatus (Persoon : Fries) Fries
Hygrophorus unguinosus (Fries : Fries) Fries
Hygrophorus unguinosus var. *subaromaticus* A. H. Smith and Hesler

COMMON NAME: none.

CAP: 2–5 cm wide, hemispheric when young, becoming convex and nearly plane in age; surface slimy when fresh, shiny when dry, dark brownish black when young, becoming brown to pale grayish brown in age; margin translucent-striate; flesh grayish to white; odor rarely faintly disagreeable or not distinctive; taste not distinctive.

GILLS: adnate or adnexed, subdistant, broad, intervenose, waxy, white to grayish white.

STALK: 3–9 cm long, 2–5 mm thick, nearly equal, slimy when fresh, pale brownish gray to grayish brown.

MICROSCOPIC FEATURES: spores 6–10 × 4–6 µm, ellipsoid, smooth, hyaline, inamyloid.

OCCURRENCE: scattered or in groups on the ground, often among mosses in conifer and broadleaf woods and the borders of swamps; July–December; widely distributed; occasional.

EDIBILITY: unknown.

OBSERVATIONS: This waxcap is so viscid and slippery that it is difficult to grasp. The combination of a slimy, brownish black cap that becomes grayish brown in age, white to grayish white gills, and a slimy, pale brownish gray to grayish brown stalk is distinctive. The name *irrigata* means "of wet places." The synonym epithet *unguinosa* means "anointed," referring to the fruitbody's thick, glutinous coating.

Hygrocybe lacmus (Schumacher) P. D. Orton and Watling Illus. p. 99

SYNONYMS:

Camarophyllus subviolaceus (Peck) Singer
Hygrocybe subviolacea (Peck) P. D. Orton and Watling
Hygrophorus subviolaceus Peck
Hygrophorus rainierensis Hesler and A. H. Smith

COMMON NAME: none.

CAP: 2–6 cm wide, broadly convex to hemispherical or broadly conical at first, becoming nearly plane and subumbonate or somewhat depressed on the disc; margin decurved and remaining so well into maturity, sometime upturned and wavy in age; surface lubricous to viscid, glabrous, hygrophanous, translucent-striate along the margin or halfway up to the center when moist, bluish gray to violet-gray or grayish violet, often becoming brownish violaceous over the disc as specimens mature; flesh thick and firm over the disc, thin on the margin, colored like the cap surface or paler; odor described as somewhat earthy, unpleasant, or not distinctive; taste variously described as unpleasant, rancid, acidic, bitter to somewhat nauseating or as acrid and at times leaving a burning in the throat.

GILLS: decurrent, arcuate, subdistant to distant, medium broad, sometimes forking, typically intervenose, pale gray to bluish gray or dull violaceous gray; edges even.

STALK: 3–7 cm long, 4–12 mm thick, nearly equal or tapered downward near the base, often curved, dry, appearing smooth, hollow in age, white to dull white or tinted like the cap especially near the apex.

MICROSCOPIC FEATURES: spores 6–8.5 × 4–7 µm, broadly ellipsoid to ovoid or sometimes subglobose, smooth, hyaline, inamyloid.

OCCURRENCE: scattered or in groups in grassy areas or in conifer or broadleaf woods, usually in wet habitats; June–November; widely distributed especially in the Northeast; uncommon.

EDIBILITY: unknown.

OBSERVATIONS: *Hygrocybe canescens* (p. 19) is similar but has a dry cap and stalk coated with tomentose patches and somewhat smaller spores. Compare also with *Hygrocybe murina* (p. 39), which is similar but has a mild odor and taste. Some authors consider *Hygrophorus rainierensis* to be a distinct species; however, the principal difference is that its flesh has a cornsilk odor, so we have placed it in synonymy with *Hygrocybe lacmus.* The name *lacmus* means "dark violet-blue."

Hygrocybe laeta var. ***laeta*** (Persoon : Fries) P. Kummer Illus. p. 125

SYNONYMS: *Hygrophorus laetus* (Persoon : Fries) Fries

COMMON NAME: none.

CAP: 1–5 cm wide, hemispheric when young, becoming convex to nearly flat in age, often depressed at the center; surface glutinous to viscid when fresh, shiny when dry, smooth, color variable, shell pink to peach, orange-red to orange to olive orange, or shades of violet-gray, yellow, or cream; margin translucent-striate; flesh thin; odor often somewhat mephitic to sweetish or not distinctive; taste not distinctive.

GILLS: adnate to decurrent, distant to subdistant, narrow, color variable like the cap.

STALK: 1.5–8 cm long, 1.5–6 mm thick, equal or slightly enlarged at the apex, hollow, glutinous to viscid when fresh, colored like the cap.

MICROSCOPIC FEATURES: spores 5–10 × 3–6 µm, ellipsoid, smooth, hyaline, inamyloid.

OCCURRENCE: scattered or in groups in sphagnum bogs or on the ground in mixed woods, often among mosses; August–December; widely distributed; common.

EDIBILITY: reported to be edible.

OBSERVATIONS: This common and widespread waxcap is highly variable in color, but the small size, exceedingly viscid cap and stalk, and oftentimes a mephitic odor are distinctive features. *Hygrocybe laeta* f. *pallida* (A. H. Smith) Bon is nearly identical except that the cap is white. The name *laeta* means "pleasing."

Hygrocybe marginata var. ***concolor*** (A. H. Smith) comb. nov. Illus. p. 104

SYNONYMS: *Hygrophorus marginatus* var. *concolor* A. H. Smith

COMMON NAME: none.

CAP: 1.5–4 cm wide, conic, becoming campanulate to nearly flat; surface smooth, moist, slightly viscid when wet, orange-yellow when young, becoming bright orange-yellow to bright golden yellow to yellow at maturity; flesh thin, yellow; odor and taste not distinctive.

GILLS: adnate, distant, broad, waxy, orange-yellow to yellow.

STALK: 2.5–10 cm long, 3–9 mm thick, nearly equal, becoming hollow in age, smooth, moist, pale yellow to orange-yellow, sometimes white at the base.

MICROSCOPIC FEATURES: spores 7–10 × 4–7 µm, ellipsoid, smooth, hyaline, inamyloid.

OCCURRENCE: single to scattered on the ground or among sphagnum mosses at the edge of bogs and in mixed woods; July–December; widely distributed but most frequently encountered in the Northeast; fairly common.

EDIBILITY: unknown.

OBSERVATIONS: *Hygrocybe marginata* var. *marginata* (p. 34) is similar, but its gills are brilliant orange and remain so in age. The name *marginata* means "having a distinct margin," but it is difficult to see the application here.

Hygrocybe marginata var. ***marginata*** (Peck) comb. nov. Illus. p. 105

SYNONYMS: *Hygrophorus marginatus* var. *marginatus* Peck

COMMON NAME: Orange-Gilled Waxcap.

CAP: 1–5 cm wide, obtusely conic at first, becoming convex to campanulate and sometimes nearly plane at maturity; surface moist, lubricous, hygrophanous, orange to yellow-orange, fading to pale yellowish or nearly white in age; margin incurved and remaining so well into maturity, sometimes faintly striate; flesh thin, fragile, colored like the cap surface; odor and taste not distinctive.

GILLS: adnexed, becoming nearly free in age, subdistant, intervenose, brilliant orange and remaining so after the cap has faded or has been washed out; edges even.

STALK: 4–10 cm long, 3–6 mm thick, nearly equal, often curved, round or compressed, moist or dry but not viscid, hollow, pale orange-yellow.

MICROSCOPIC FEATURES: spores 7–10 × 4–6 µm, ellipsoid to suboblong, smooth, hyaline, inamyloid.

OCCURRENCE: solitary, scattered, or in groups on humus and soil in mixed woods; June–December; widely distributed; fairly common.

EDIBILITY: reported to be edible, but of little culinary interest.

OBSERVATIONS: *Hygrocybe marginata* var. *concolor* (p. 33) is similar, but it has an orange-yellow to bright golden yellow cap, and its gills are also orange-yellow. *Hygrocybe marginata* var. *olivacea* (p. 35) has an olive brown to dark olive cap when young that develops pale dull orange coloration near the margin at maturity and ochraceous orange gills. The variety name *concolor* refers to the matching cap and gill color.

Hygrocybe marginata var. ***olivacea*** (A. H. Smith and Hesler) comb. nov.
Illus. p. 105

SYNONYMS:
Hygrophorus marginatus var. *olivaceus* A. H. Smith and Hesler
Tricholoma marginatum var. *olivaceum* (A. H. Smith and Hesler) Singer

COMMON NAME: none.

CAP: 2.5–6 cm wide, conic at first, becoming conic-campanulate and finally nearly plane, often with a sharp conic umbo; surface glabrous, moist but not viscid, olive brown to deep olive in the center, usually with dull orange tones near the margin especially in older specimens; margin translucent-striate; flesh thin and fragile, dull olive gray; odor and taste not distinctive.

GILLS: adnexed, often with a decurrent tooth, broad, close to subdistant, orange and scarcely fading in age.

STALK: 3–5 cm long, 2–4 mm thick, nearly equal overall, soon hollow, very fragile, glabrous, pale green when very young, soon becoming pale greenish yellow to pale orange-yellow to yellow, often whitish toward the base.

MICROSCOPIC FEATURES: spores 6.5–8 × 4–5 µm, subellipsoid, smooth, hyaline, inamyloid.

OCCURRENCE: solitary, scattered, or in groups on the ground or on well-decayed conifer wood; June–November; widely distributed in the Northeast; uncommon.

EDIBILITY: unknown.

OBSERVATIONS: Note the striking contrast between the olive brown to olive colors on the cap and the orange gills. *Hygrocybe marginata* var. *concolor* (p. 33) and *H. marginata* var. *marginata* (p. 34) have orange or yellow-orange to yellow caps that lack olive tones. The variety name *olivacea* refers to the cap's olivaceous tones.

Hygrocybe mephitica (Peck) Courtecuisse Not Illustrated

SYNONYMS: *Hygrophorus mephiticus* Peck

COMMON NAME: none.

CAP: 2–4 cm wide, convex, becoming nearly plane; margin somewhat striate when moist; surface hygrophanous, yellowish brown when moist, ochraceous when dry, sometimes tinged green; flesh whitish, sometimes tinged yellow; odor mephitic; taste not reported.

GILLS: adnexed and sinuate, distant, sometimes intervenose, grayish violaceous or grayish purple.

STALK: 3–5 cm long, 2–5 mm thick, nearly equal or tapered downward, sometimes curved, moist or dry, hollow, brittle, concolorous with the cap or paler; base often coated with a white mycelium.

MICROSCOPIC FEATURES: spores 8.5–12 × 5–7 µm, ellipsoid, smooth, hyaline, inamyloid.

OCCURRENCE: scattered or in groups among sphagnum mosses in swamps; August; Massachusetts, distribution limits and frequency yet to be determined.

EDIBILITY: unknown.

OBSERVATIONS: The name *mephitica* means "mephitic," a reference to the odor of the flesh. The flesh of *Hygrocybe auratocephala* (p. 16) also has a mephitic odor, but this variety has yellow gills and smaller spores that measure 7–9 × 4–6 µm. Compare with *Hygrocybe purpureofolia* (p. 47), which is somewhat similar but has a dark orange-red to bright yellowish orange cap, adnate to slightly decurrent lavender to purple gills that fade in age, and flesh that lacks a distinctive odor.

Hygrocybe miniata var. ***miniata*** (Fries) Kummer Illus. p. 115

SYNONYMS:

Hygrocybe constans Murrill
Hygrocybe flammea (Scopoli) Murrill
Hygrocybe miniata var. *mollis* (Berkeley and Broome) Arnolds
Hygrophorus congelatus Peck
Hygrophorus miniatus (Fries) Fries
Hygrophorus miniatus var. *typicus* A. H. Smith and Hesler

COMMON NAME: Fading Scarlet Waxcap.

CAP: 2–4 cm wide, convex, becoming broadly convex to nearly flat with a depressed disc in age; surface smooth and brilliant scarlet when moist, fading to orange or yellow, becoming dry and fibrillose-scurfy; flesh thin, colored like the cap or paler; odor and taste not distinctive.

GILLS: adnate, sometimes slightly decurrent, close to subdistant, broad, colored like the cap when young, becoming paler orange, yellow, or orange-pink in age.

STALK: 2.5–5 cm long (up to 7 cm when growing in deep humus or among mosses), 3–6 mm thick, dry, equal or compressed, stuffed, smooth, colored like the cap when young, fading to orange-yellow or yellow in age.

MICROSCOPIC FEATURES: spores 6–8 × 4–6 µm, ellipsoid, often constricted in the middle and widened near the base, uninucleate, smooth, hyaline, inamyloid.

OCCURRENCE: scattered or in groups on the ground, among mosses, or on decaying wood in broadleaf and mixed woods; July–January; widely distributed; fairly common.

EDIBILITY: reported to be edible.

OBSERVATIONS: This highly variable species has been the subject of much confusion among mycologists, who have provided an array of descriptions and names. We have placed *Hygrocybe miniata* var. *mollis* in synonymy because of insufficient differentiating characteristics. *Hygrocybe miniata*

f. *longipes* (A. H. Smith and Hesler) comb. nov. (p. 115) is nearly identical but usually has a much longer stalk and a distinctly translucent-striate cap that becomes slightly squamulose in age. *Hygrocybe calciphila* (p. 18) is nearly identical but has broadly ellipsoid binucleated spores that are not constricted near the middle or widened near the base. The name *miniata* refers to the color red lead or cinnabar red.

Hygrocybe minutula (Peck) Murrill — Illus. p. 114

SYNONYMS: *Hygrophorus minutulus* Peck

COMMON NAME: none.

CAP: 5–15 mm wide, convex, becoming broadly convex to nearly plane; margin decurved and remaining so well into maturity, somewhat striate when fresh; surface distinctly viscid or glutinous, glabrous, hygrophanous, scarlet to reddish orange, fading to orange-yellow or yellow in age; flesh thin, fragile, colored like the cap surface, not staining when bruised; odor and taste not distinctive.

GILLS: adnate to adnexed with a decurrent tooth, close to subdistant, pale orange to yellowish orange or pale orange-yellow; edges even.

STALK: 1.5–5 cm long, 1–3 mm thick, nearly equal overall or tapered downward, sometimes constricted, hollow in age, fragile, glabrous, viscid or glutinous, reddish nearly overall or on the upper portion, yellowish to whitish on the lower portion, fading to yellow overall in age.

MICROSCOPIC FEATURES: spores 7–10 × 4–6 µm, ellipsoid, smooth, hyaline, inamyloid.

OCCURRENCE: scattered or in groups in grassy areas or on the ground, usually in broadleaf woodlands; widely distributed; occasional.

EDIBILITY: unknown.

OBSERVATIONS: The very small cap and stalk size, the viscid to glutinous reddish cap and stalk, and lack of staining black on any parts separate this mushroom from most other similar species. *Hygrocybe subminutula* (Murrill) Pegler (p. 116), reported from Florida, is nearly identical except that its stalk apex is quite persistently red, and it has smaller and narrower spores that measure 5–7 × 2.5–3.5 µm. The name *minutula* refers to this waxcap's small size.

Hygrocybe mississippiensis (Cibula) comb. nov. — Illus. p. 116

SYNONYMS: *Hygrophorus mississippiensis* Cibula

COMMON NAME: Mississippi Waxcap.

CAP: 5–10 mm wide, convex to nearly plane, becoming broadly depressed at times; surface dry, appearing squamulose over the disc, bright red to deep red, paler near the margin; margin sometimes becomes white or yellowish with age; flesh thin, reddish, hygrophanous; odor and taste not distinctive.

GILLS: adnexed with a decurrent tooth, sometimes emarginate at maturity, subdistant, pale yellow to light yellowish brown, at times yellowish pink.

STALK: 2–5 cm long, 1–3 mm thick, nearly equal, glabrous with a silky luster, hollow, colored like the cap but lighter, especially at the apex, darker below.

MICROSCOPIC FEATURES: spores 7.5–8.4 × 4.7–5.8 µm, ellipsoid to subovate, smooth, hyaline, inamyloid.

OCCURRENCE: gregarious on leaf litter, soil, and decaying stumps in mixed bottomland forests; July–September; reported from Mississippi but to be expected in similar habitats in the Gulf Coast region; locally common.

EDIBILITY: unknown.

OBSERVATIONS: *Hygrocybe mexicana* Singer, reported from Mexico, has a bright scarlet red, glabrous cap, broadly adnate yellow gills mixed with pinkish red, and a bright scarlet red stalk. *Hygrocybe firma* var. *trinitensis* Dennis, reported from Trinidad, is similar but has a scarlet cap with a narrow yellow zone around the margin, decurrent coral red gills, and a scarlet stalk with a yellowish base. The species name refers to the state of Mississippi, where this waxcap was first described.

Hygrocybe mucronella (Fries) P. Karsten Illus. p. 111

SYNONYMS:

Hygrocybe reai (Maire) J. E. Lange
Hygrophorus reai Maire

COMMON NAME: Bitter Waxcap.

CAP: 1–3.5 cm wide, hemispherical or convex to campanulate, becoming broadly conical to nearly flat; surface smooth, lubricous when young but soon dry, brilliant orange-red on the disc, gradually fading to reddish orange to bright orange and yellow-orange toward the margin; margin slightly translucent-striate and sometimes crenate; flesh thin, reddish orange; odor not distinctive; taste very bitter.

GILLS: adnate to short decurrent with a tooth, subdistant, broad, waxy, usually orange with a paler edge, sometimes salmon tinged, becoming more orange-yellow in age.

STALK: 1.5–6.3 cm long, 1.5–6 mm thick, nearly equal or tapered downward near the base, cylindrical or sometimes compressed, smooth, moist to viscid when young but soon dry, colored like the cap or paler orange or yellow, occasionally with a whitish base.

MICROSCOPIC FEATURES: spores 6–10 × 4–7 µm, irregularly ellipsoid to oblong, often constricted in the middle or widened near the base, smooth, hyaline, inamyloid.

OCCURRENCE: scattered or in groups usually under conifers, in grassy areas, and often among mosses; July–October; widely distributed in the Northeast; occasional.

EDIBILITY: inedible.

OBSERVATIONS: The bitter taste of the cap surface and flesh distinguishes this species from the nearly identical *Hygrocybe minutula* (p. 37), which lacks the bitter taste. The name *mucronella* is derived from the Latin word *mucron*, meaning "sharply pointed."

Hygrocybe murina (Bird and Grund) comb. nov. Not Illustrated

SYNONYMS: *Hygrophorus murinus* Bird and Grund

COMMON NAME: none.

CAP: 2–6 cm wide, convex to flat, often subumbonate, becoming shallowly depressed with an uplifted margin; surface viscid to resinous, zonate at times, glabrous to appressed-fibrillose, dark gray-brown when young, becoming pale cream in age; flesh thick, becoming thinner in age, whitish, gray near the cap cuticle; odor and taste not distinctive.

GILLS: decurrent, subdistant, narrow, often intervenose, gray, becoming yellowish with age; edges even.

STALK: 3–5 cm long, 4–12 mm thick, equal or tapering down, appressed-fibrillose, white, tinged cream, at times with a yellowish base, especially after being handled.

MICROSCOPIC FEATURES: spores 7–10 × 5–6 µm, ellipsoid, smooth, hyaline, inamyloid.

OCCURRENCE: gregarious to clustered under hemlock and white birch; October–November; reported only from Nova Scotia; uncommon.

EDIBILITY: unknown.

OBSERVATIONS: The mild odor and taste help distinguish this species from *Hygrocybe lacmus* (p. 32), which has an unpleasant taste and at times an unpleasant odor, and *H. nordmanensis* (Hesler and A. H. Smith) comb. nov., which occurs in the western United States and has the odor of green corn. The name *murina* means "mouse gray."

Hygrocybe mycenoides (A. H. Smith and Hesler) Lodge and S. A. Cantrell Not Illustrated

SYNONYMS: *Hygrophorus mycenoides* A. H. Smith and Hesler

COMMON NAME: none.

CAP: 6–12 mm wide, convex or conical at first, becoming convex to broadly convex or nearly plane with a slight umbo or slightly depressed disc at maturity; margin incurved when young then elevated in age, translucent-striate to the disc when moist; surface moist and lubricous but not viscid, deep yellow-orange to orange-yellow over the center, paler yellow toward the margin, fading to nearly whitish in age; flesh thin, waxy, fragile, bright yellow fading to white; odor and taste not distinctive.

GILLS: broadly adnate to subdecurrent, subdistant to close, narrow, bright yellow, the faces fading before the edges and thus appearing marginate, whitish overall in age, not staining when bruised.

STALK: 2.5–4 cm long, 1–1.5 mm thick, nearly equal, slender, hollow at maturity, pale yellow or colored like the cap surface.

MICROSCOPIC FEATURES: spores 5–7 × 2.5–3.5 µm, narrowly ellipsoid, smooth, hyaline, inamyloid.

OCCURRENCE: scattered on humus in mixed woodlands; August; reported from Tennessee; seasonal range, distribution, and frequency yet to be determined.

EDIBILITY: unknown.

OBSERVATIONS: The very small stature and color combination of its cap, stalk, and gills make this mushroom distinctive. *Hygrocybe parvula* (p. 43) is similar but has a larger cap, whitish to pale yellow intervenose gills, and a stalk that is often darker colored than its cap surface. The name *mycenoides* refers to the stature, which resembles that of gilled mushrooms in the genus *Mycena*.

Hygrocybe nitida (Berkeley and M. A. Curtis) Murrill Illus. p. 102

SYNONYMS: *Hygrophorus nitidus* Berkeley and M. A. Curtis

COMMON NAME: none.

CAP: 1–4 cm wide, broadly convex or flattened when young, disc very soon becoming depressed, in age deeply infundibuliform, sometimes perforating; margin incurved and remaining so well into maturity; surface smooth, glabrous, viscid, somewhat striate when moist, bright yellow to apricot yellow, fading to pale yellow or whitish; flesh soft, fragile, very thin, yellowish to whitish; odor and taste not distinctive.

GILLS: long-decurrent, subdistant to distant, yellow to pale yellow, often with darker yellow edges, intervenose; edges even.

STALK: 3–8 cm long, 2–5 mm thick, nearly equal or slightly enlarged in the upper portion, smooth, viscid, colored and fading like the cap, hollow.

MICROSCOPIC FEATURES: spores 6.5–9 × 4–6 µm, ellipsoid to subovoid, smooth, hyaline, inamyloid.

OCCURRENCE: scattered or in groups on humus, in wet soil, or among mosses in conifer or broadleaf forests and in bogs; June–December; widely distributed; occasional to fairly common.

EDIBILITY: unknown.

OBSERVATIONS: Compare with *Hygrocybe cantharellus* f. *cantharellus* (p. 20), which is similar but has a dry, not viscid, orange to orange-red or scarlet cap and stalk. The name *nitida* means "glossy," in reference to the shiny aspect of this waxcap.

Hygrocybe nitrata (Persoon) Wünsche Illus. p. 91

SYNONYMS:

Camarophyllus nitratus (Persoon) Ricken
Hygrophorus nitratus Fries

COMMON NAME: none.

CAP: 2–7 cm wide, campanulate when young, becoming convex in age; surface dry, not viscid, smooth on young specimens then breaking up into fine squamules and fibers in older specimens, gray-brown to grayish brown, often with ivory or buff streaks, becoming dark grayish brown to blackish in old specimens; flesh thin, pale to dark brownish gray, not staining when cut and rubbed; odor pungent, medicinal, strongly nitrous; taste acidic.

GILLS: adnexed to sinuate or sometimes nearly free in age, subdistant, broad, waxy, typically intervenose, ivory with grayish tints and darkening in age, not staining when bruised.

STALK: 4–10 cm long, 2–10 mm thick, nearly equal or enlarged at the base, smooth, hollow, pale brown to ivory brown when young, becoming grayish brown in age, whitish at the base.

MICROSCOPIC FEATURES: spores 7.5–10 × 4.5–6 µm, oval to ellipsoid, smooth, hyaline, inamyloid.

OCCURRENCE: scattered or in groups on the ground under conifers and broadleaf trees; July–December; widely distributed; occasional.

EDIBILITY: unknown.

OBSERVATIONS: The combination of the nitrous odor and the gray-brown colors distinguishes this mushroom from similar species. *Hygrocybe ovina* (p. 42) is similar, but it has a much darker cap and stalk; its flesh and gills readily stain pinkish to reddish; and its flesh typically lacks the pronounced nitrous odor. Some authors consider *Hygrocybe murinacea* (Bulliard : Fries) Moser to be a synonym of *H. nitrata*. The name *nitrata* refers to the nitrous (ammonia-like) odor of the flesh.

Hygrocybe occidentalis var. ***occidentalis*** (Dennis) Pegler Illus. p. 97

SYNONYMS: *Hygrophorus occidentalis* A. H. Smith and Hesler

COMMON NAME: none.

CAP: 2–10 cm wide, convex, soon becoming nearly plane, disc slightly depressed at maturity; margin decurved and remaining so well into maturity; surface viscid, typically glutinous, somewhat appressed-fibrillose, streaked, color variable, brown to fuscous or sometimes grayish or yellowish over the disc, whitish to grayish toward the margin; flesh thin, white; odor and taste not distinctive.

GILLS: adnate at first, becoming subdecurrent or decurrent, close to subdistant, narrow to moderately broad, thin, white or tinged cream; edges even.

STALK: 2–10 cm long, 3–15 mm thick, nearly equal down to a tapered base, solid, often minutely punctate or pruinose near the apex, viscid or glutinous, whitish or colored like the cap surface.

MICROSCOPIC FEATURES: spores 6–8 × 3.5–5 µm, ellipsoid, smooth, hyaline, inamyloid.

OCCURRENCE: scattered or in groups on the ground in broadleaf or mixed woodlands, often associated with oak or hickory; widely distributed; occasional.

EDIBILITY: unknown.

OBSERVATIONS: The name *occidentalis* means "western," in reference to western North America, where this waxcap is fairly common.

Hygrocybe ovina (Bulliard) Kühner Illus. p. 95

SYNONYMS:

Hygrocybe ingrata Jensen and Möeller
Hygrophorus nitiosus Blytt
Hygrocybe nitiosa (Blytt) Moser

COMMON NAME: none.

CAP: 2–8 cm wide, hemispheric to convex at first, becoming broadly convex at maturity, sometimes with a low broad umbo or shallowly depressed on the disc; margin even, incurved and remaining so well into maturity, becoming decurved in age; surface dry, smooth at first, becoming radially appressed-fibrillose, sometimes uneven and cracked, dark brown to dark grayish brown or sometimes paler when young, at times with reddish brown or olivaceous tones; flesh moderately thick, whitish to pale grayish brown, slowly staining reddish when exposed; odor fruity to faintly nitrous or not distinctive; taste acidic.

GILLS: adnexed at first, becoming deeply emarginate in age, subdistant, thick, broad, whitish to cream at first, developing grayish or brownish tints in age, staining reddish and sometimes slowly blackish when bruised; edges even.

STALK: 4–9 cm long, 4–15 (20) mm thick, nearly equal or enlarged downward, sometimes fusiform, often furrowed or compressed, dry, stuffed to hollow, glabrous, colored like the cap or pale brown with a whitish base, sometimes darkening when handled or in age.

MICROSCOPIC FEATURES: spores 6–10 × 4.5–7 µm, ellipsoid to subovoid, smooth, hyaline, inamyloid.

OCCURRENCE: sometimes solitary but more often in groups on soil in grassy places or in broadleaf woods; July–December; widely distributed; uncommon.

EDIBILITY: unknown.

OBSERVATIONS: This uncommon waxcap has been confused in the literature because of the variation in color and odor of the flesh, which has caused some authors to treat *Hygrocybe nitiosa* as a separate taxon. *Hygrocybe nitrata* (p. 41) is similar, and its flesh has a nitrous odor, but it has a grayish brown cap, and its gills and flesh do not stain reddish when cut or bruised. *Hygrocybe subovina* (p. 53) is also similar but has a darker brown cap, globose to subglobose spores, and whitish to brownish nonstaining flesh with a fragrant odor. The name *ovina* refers to sheep, in reference either to the pastures where this mushroom may occur or to the nitrous odor that is typical where sheep are present.

Hygrocybe parvula comb. nov. Illus. p. 101

SYNONYMS: *Hygrophorus parvulus* Peck

COMMON NAME: none.

CAP: 1–3 cm wide, convex to broadly convex or nearly flat, sometimes depressed on the disc; surface smooth, viscid, translucent-striate, orange-yellow or amber yellow, becoming paler yellow in age; flesh waxy, colored like the cap; odor and taste not distinctive.

GILLS: decurrent, subdistant to distant, intervenose, whitish to cream or pale yellow.

STALK: 3–6 cm long, 1.5–3 mm thick, equal or tapered at the base, hollow, fragile, smooth, not viscid, colored like the cap, lower portion and base sometimes darker pinkish red to reddish brown or yellow-brown.

MICROSCOPIC FEATURES: spores 5–8.5 × 3.5–5 µm, ellipsoid, smooth, hyaline, inamyloid.

OCCURRENCE: scattered on the ground under broadleaf or mixed woods and under rhododendron; July–November; widely distributed but more common in the Northeast; occasional.

EDIBILITY: unknown.

OBSERVATIONS: As Peck pointed out in his original description, the stalk is often more highly colored than the cap. Compare with *Hygrocybe flavescens* (p. 28), which has a larger cap and adnexed gills that do not become decurrent. The name *parvula* means "very small."

Hygrocybe pratensis var. ***pratensis*** (Persoon : Fries) Murrill Illus. p. 87

SYNONYMS:

Camarophyllus pratensis (Fries) Kummer

Hygrophorus pratensis var. *pratensis* (Persoon : Fries) Fries

COMMON NAME: Salmon Waxcap, Meadow Waxcap.

CAP: 2–7 cm wide, obtuse to convex when young, becoming broadly convex, typically with a low broad umbo at maturity; surface smooth, moist then dry, often cracked on the disc in age, reddish orange to salmon orange or dull orange when young, fading to pale orange to orange-yellow or pale tawny in age; margin incurved at first, sometimes wavy at maturity; flesh thick, brittle, whitish to pale reddish cinnamon; odor and taste not distinctive.

GILLS: decurrent, subdistant to distant, often intervenose, salmon buff to pale orange.

STALK: 3–8 cm long, 5–20 mm thick, usually tapered downward or nearly equal, often curved, dry, smooth, whitish to pale salmon buff.

MICROSCOPIC FEATURES: spores 5–8 × 3–5 µm, ellipsoid, subovoid, or subglobose, smooth, hyaline, inamyloid.

OCCURRENCE: solitary, scattered, or in groups in grassy areas or on humus in woods; May–January (in Deep South); widely distributed; occasional.

EDIBILITY: edible.

OBSERVATIONS: *Hygrocybe pratensis* var. *robusta* (p. 44) is similar but is more robust, having a pinkish cinnamon to dull pink cap, whitish and unevenly decurrent gills with pinkish tinges, and a whitish, firm, deeply rooting stalk that has pinkish tinges and is conspicuously narrowed near the base. *Hygrocybe pratensis* var. *pallida* (Cooke) Arnolds (p. 85) is a variety reported from Europe and North America that has a white cap and white to ivory or pale buff gills. The name *pratensis* means "of meadows."

Hygrocybe pratensis var. ***robusta*** (Hesler and A. H. Smith) comb. nov.

Not Illustrated

SYNONYMS: *Hygrophorus pratensis* var. *robustus* Hesler and A. H. Smith

COMMON NAME: none.

CAP: 4–10 cm wide, obtuse to convex, sometimes with a low broad umbo; surface dull and unpolished, often uneven, sometimes finely cracked, dry, pinkish cinnamon to light pinkish cinnamon or dull pink, staining yellowish and finally brownish along the margin when injured; margin spreading and elevated to upturned or sometimes remaining decurved, lobed to crenate in age; flesh thick, firm, pale pinkish buff, sometimes staining yellowish when cut, not reacting with KOH or $FeSO_4$, odor and taste not distinctive.

GILLS: decurrent and often unequally so, subdistant to distant, forked, thick, broad, light pink to pale pinkish cinnamon; edges even.

STALK: 4–12 cm long, 1–3 cm thick, tapered downward to a conspicuously narrowed and deeply rooting base, solid, firm, dry, whitish and unpolished at first, developing pinkish buff streaks and tinges at maturity, pale pinkish buff within.

MICROSCOPIC FEATURES: spores 7–8 × 5–5.5 µm, ellipsoid, smooth, hyaline, inamyloid.

OCCURRENCE: scattered or in groups on the ground or among mosses in broadleaf woodlands; August–November; widely distributed; uncommon.

EDIBILITY: edible.

OBSERVATIONS: *Hygrocybe pratensis* var. *pratensis* (p. 43) is similar but has a reddish orange to salmon orange or dull orange cap, salmon buff to pale orange gills, and a whitish to pale salmon buff stalk. *Hygrocybe pratensis* var. *pallida* (Cooke) Arnolds (p. 85) has a white cap and white to ivory or pale buff gills. *Hygrophorus pudorinus* var. *pudorinus* f. *pudorinus* (p. 72) has a viscid, pale pinkish buff to pale flesh-colored cap and white to whitish subdecurrent gills that are sometimes tinged pinkish, and it grows on the ground or among mosses in conifer woods. The variety name *robusta* refers to this form's stout fruitbody.

Hygrocybe psittacina var. ***perplexa*** (A. H. Smith and Hesler) Boertmann
Illus. p. 127

SYNONYMS: *Hygrophorus perplexus* A. H. Smith and Hesler
COMMON NAME: Perplexing Waxcap.
CAP: 1–3 cm wide; conic, becoming nearly flat, with a broad umbo; surface smooth, slimy, color variable, orange-brown to pinkish brown or brownish orange to reddish orange when young, becoming dull orange to orange-yellow, sometimes with an olive tint; margin incurved when young, translucent-striate when fresh; flesh colored like the cap; odor and taste not distinctive.
GILLS: adnate, subdistant, ivory yellow with faint pink tones when young, becoming yellow to orange-yellow or apricot in age.
STALK: 3–5 cm long, 2–5 mm thick, nearly equal or enlarged slightly downward, hollow, smooth, slimy, color variable from concolorous with the cap to amber gold to pale yellow-tan, sometimes with faint pinkish gray or olivaceous tints toward the apex.
MICROSCOPIC FEATURES: spores 6–8 × 4–5 µm, oval to ellipsoid, smooth hyaline, inamyloid.
OCCURRENCE: in groups or in clusters on the ground under broadleaf trees, especially beech; July–February; widely distributed; occasional.
EDIBILITY: unknown.
OBSERVATIONS: *Hygrocybe psittacina* var. *psittacina* (p. 45) is nearly identical, but its cap colors are various shades of yellow, green, olive brown, pink, or bluish green; its stalk typically has green near the apex; and it has slightly larger spores. The name *perplexa* refers to the difficulty in distinguishing this variety from *H. psittacina* var. *psittacina*.

Hygrocybe psittacina var. ***psittacina*** (Schaeffer : Fries) P. Kummer
Illus. p. 128

SYNONYMS: *Hygrophorus psittacinus* (Schaeffer : Fries) Fries
COMMON NAME: Parrot Mushroom.
CAP: 1–3 cm wide; broadly conic, becoming convex to nearly flat, sometimes

with a broad umbo; surface viscid when fresh, smooth and shiny when dry, typically dark green at first then variously colored, often a mixture of yellow, green to olive green, amber orange, orange, pinkish orange, pinkish, olive brown, or bluish green; margin translucent-striate at first, becoming opaque in age; flesh thin, colored like the cap; odor and taste not distinctive.

GILLS: adnate to notched, subdistant, waxy, pale yellow-orange, yellow, greenish, or reddish.

STALK: 3–7 cm long, 2–5 mm thick, smooth, viscid overall, nearly equal or tapered slightly upward, hollow, color variable, green when young, soon becoming amber orange, yellow, pinkish or yellowish white, often retaining green to olive tints, especially near the apex.

MICROSCOPIC FEATURES: spores 6.5–10 × 4–6 µm, ellipsoid, smooth, hyaline, inamyloid.

OCCURRENCE: scattered or in groups on the ground under conifers and broadleaf trees and in grassy areas; June–February; widely distributed; fairly common but easily overlooked, especially when young.

EDIBILITY: reported to be edible.

OBSERVATIONS: Compare with *Hygrocybe psittacina* var. *perplexa* (p. 45). Compare also with *Hygrocybe laeta* var. *laeta* (p. 33), which has duller cap colors and typically a mephitic or burned rubber odor. The Parrot Waxcap is one of a handful of small hygrocybes that are so viscid that they are nearly impossible to pick up. The name *psittacina* refers to the parrot family (Psittacidae) because of this variety's multicolored fruit bodies.

Hygrocybe punicea (Fries : Fries) Kummer — Illus. p. 116

SYNONYMS: *Hygrophorus puniceus* Fries

COMMON NAME: Crimson Waxcap.

CAP: 2–12 cm wide, obtusely conic, becoming broadly conic to nearly plane, often with a conspicuous umbo; margin incurved at first, becoming decurved to uplifted, sometimes splitting; surface glabrous, viscid to lubricous, deep blood red to dark red overall at first, fading to dull orange and finally yellowish to buff or a mixture of these colors; flesh thin, fragile, waxy, pale reddish orange to pale yellow, often white near the center, not staining when exposed; odor and taste not distinctive.

GILLS: narrowly adnate at first, becoming adnexed to emarginate with a decurrent tooth, subdistant, broad, reddish orange to orange-yellow or pale yellow; edges even and yellowish.

STALK: 2–12 cm long, 5–15 mm thick, nearly equal overall, fusiform, or tapered in either direction, sometimes compressed, hollow in age, dry, fibrillose-striate, usually variously colored red to orange-red, orange or yellow, with a white to yellowish base.

MICROSCOPIC FEATURES: spores 7–12 × 3.5–6 µm, oblong to ellipsoid, often constricted near the middle, smooth, hyaline, inamyloid.

OCCURRENCE: scattered or in groups in grassy areas or on the ground under conifers and broadleaf trees; June–December; widely distributed; fairly common.

EDIBILITY: reported to be edible.

OBSERVATIONS: Compare with *Hygrocybe coccinea* (p. 23), which typically has a smaller cap that is moist to lubricous but not viscid, is bright red when young, and has red to reddish orange flesh, broadly adnate gills, and a smooth stalk. Because size and fruitbody colors are not entirely reliable features in separating *Hygrocybe punicea* and *Hygrocybe coccinea,* these two species can be difficult to distinguish in the field. The name *punicea* means "crimson."

Hygrocybe pura (Peck) comb. nov. Not Illustrated

SYNONYMS: *Hygrophorus purus* Peck

COMMON NAME: none.

CAP: 3–8 cm wide, conic to conic-campanulate, margin recurved in age; surface viscid, glabrous, white, bruising pinkish to red; flesh thin, white, waxy; odor and taste not distinctive.

GILLS: adnexed with a decurrent hook, subdistant, broad, ventricose, white.

STALK: 4–8 cm long, 3–8 mm thick, equal, hollow, glutinous, glabrous, shiny white, base bruising pinkish red.

MICROSCOPIC FEATURES: spores 7–9 × 4–5 µm, ellipsoid, smooth, hyaline, inamyloid.

OCCURRENCE: scattered in deep humus in mixed woods; August–October; New York to Alabama; uncommon.

EDIBILITY: unknown.

OBSERVATIONS: This mushroom resembles *Hygrocybe calyptriformis* (p. 19) in form but is white rather than pink. The name *pura* means "pure" or "unblemished."

Hygrocybe purpureofolia (Bigelow) Courtecuisse Illus. p. 108

SYNONYMS: *Hygrophorus purpureofolius* Bigelow

COMMON NAME: Lavender-Gilled Waxcap.

CAP: 1.2–5 cm wide, conic, becoming broadly convex to nearly plane with a broad umbo; surface smooth, moist, not viscid, dark orange-red overall, fading to bright yellowish orange in age, palest along the margin; flesh thin, white; odor and taste not distinctive.

GILLS: adnate to slightly decurrent, close to subdistant, waxy, lavender to purple, fading to orange-lavender or yellowish in age.

STALK: 2.5–7.5 cm long, 5–12 mm thick, nearly equal or enlarged downward, hollow, smooth, often compressed with a vertical groove, orange-red to orange-yellow.

MICROSCOPIC FEATURES: spores 7–11 × 4–5.5 µm, ellipsoid, smooth, hyaline, inamyloid.

OCCURRENCE: scattered or in groups on the ground in broadleaf and mixed woods; July–November; widely distributed but more common in the Northeast; occasional.

EDIBILITY: unknown.

OBSERVATIONS: The combination of a dark orange-red cap and lavender to purple gills is unusual and quite striking. Compare with *Hygrocybe mephitica* (p. 35), which has grayish violaceous to grayish purple sinuate gills, a yellowish brown cap, and a mephitic odor. The name *purpureofolia* means "purplish leaves," in reference to the color of the gills.

Hygrocybe recurvata (Peck) comb. nov. Not Illustrated

SYNONYMS:

Camarophyllus recurvatus (Peck) Murrill
Hygrophorus recurvatus Peck

COMMON NAME: none.

CAP: 1–3 cm wide, convex to nearly plane, margin wavy or pleated, sometimes recurved, center may be depressed and have a small umbo; surface dry or lubricous to slightly viscid in wet conditions, smooth or with small scales, faintly translucent-striate when moist, dark to pale olive brown, paler at the margin; flesh thin, fragile, dark olive brown; odor and taste not distinctive.

GILLS: decurrent, distant to subdistant, broad, grayish white.

STALK: 2–5 cm long, 3–6 mm thick, smooth, faintly striate at times, moist, concolorous with the cap surface or whitish, interior gray.

MICROSCOPIC FEATURES: spores 7–12 × 4–6 µm, ellipsoid, smooth, hyaline, inamyloid.

OCCURRENCE: scattered to gregarious under conifers and in grassy areas; October–January; Michigan and New York, south to Florida, also reported from northern California; uncommon in eastern North America.

EDIBILITY: unknown.

OBSERVATIONS: Collections from shaded areas tend to have paler caps and stalks than those from more exposed habitats. Compare with *Hygrocybe colemanniana* (p. 24), which is similar but has a reddish brown cap. The name *recurvata* means curved backward, in reference to the cap margin, which may become recurved.

Hygrocybe reidii Kühner Illus. p. 113

SYNONYMS: none.

COMMON NAME: none.

CAP: 1–5 cm wide, hemispherical at first, becoming convex to broadly convex or nearly plane, sometimes slightly depressed over the disc; margin often crenulate, incurved at first and remaining so for a long time, at times becoming uplifted in age; surface moist or dry but not viscid, not striate or only short-striate near the margin, hygrophanous, somewhat tomentose and cracking on drying but appearing smooth when moist, scarlet to orange-red with a narrow yellow-orange margin at first, gradually fading to orange or orange-yellow in age; flesh fairly thick, firm, colored like the cap surface or pale yellow; odor sweet, resembling honey, sometimes weak and only noticeable when rubbed or when drying; taste not distinctive.

GILLS: broadly adnate with a decurrent tooth, distant, fairly broad, thick, pale orange to dull yellow or pinkish orange, typically with a paler margin, not staining when exposed.

STALK: 2–7 cm long, 2–8 mm thick, nearly equal overall or tapered downward, sometimes compressed, smooth, dry, orange on the upper portion, pale orange to orange-yellow on the lower portion, sometimes with a white tomentum at the base.

MICROSCOPIC FEATURES: 5.5–9.5 × 3.5–5 µm, ellipsoid to obovoid, smooth, hyaline, inamyloid.

OCCURRENCE: solitary, scattered, or in groups on the ground in broadleaf woods or among sphagnum mosses; June–October; widely distributed in the Northeast; fairly common.

EDIBILITY: unknown.

OBSERVATIONS: *Hygrocybe miniata* var. *miniata* (p. 36) is similar but has a more fibrillose-scurfy cap surface, close to subdistant gills, a stalk that is scarlet at first and fades to orange-yellow or yellow, and flesh that lacks the sweetish odor. The specific epithet honors mycologist D. A. Reid.

Hygrocybe russocoriacea (Berkeley and T. K. Miller) P. D. Orton and Watling Illus. p. 85

SYNONYMS:

Camarophyllus russocoriaceus (Berkeley and T. K. Miller) J. E. Lange
Hygrophorus lawrencei Hesler and A. H. Smith
Hygrophorus russocoriaceus Berkeley and T. K. Miller in Berkeley and Broome

COMMON NAME: Cedarwood Waxcap, Russian Leather Waxcap.

CAP: 1–3 cm wide, at first hemispherical, becoming broadly convex to nearly plane, sometimes slightly umbonate or depressed; margin incurved at first then somewhat elevated, entire or sometimes split; surface slightly lubricous when moist, then dry, translucent-striate, hygrophanous, ivory white to pale buff or yellowish buff, often paler in age; flesh thin, colored like the cap surface or paler; odor strongly of cedarwood; taste medicinal to unpleasant and somewhat like cedarwood.

GILLS: short to deeply decurrent, subdistant to distant, sometimes venose or intervenose, whitish, not staining when bruised.

STALK: 2–6 cm long, 2–8 mm thick, nearly equal overall or tapered downward, often curved, dry, glabrous, colored like the cap surface or with tints of pale gray.

MICROSCOPIC FEATURES: spores 6–9 × 4.5–6 µm, ellipsoid to ovoid or oblong, smooth, hyaline, inamyloid.

OCCURRENCE: solitary, scattered, or in groups among mosses or on duff in conifer woods; July–November; New York, distribution limits and frequency yet to be determined.

EDIBILITY: unknown.

OBSERVATIONS: The pronounced odor of cedarwood makes this waxcap easy to identify. *Hygrocybe virginea* var. *virginea* (p. 54) is similar but has a somewhat whiter cap and lacks a distinctive odor. Compare *Hygrocybe russocoriacea* also with *Hygrophorus piceae* (p. 71), which is similar in stature but lacks a distinctive odor. The name *russocoriacea* means "Russian leather," which has an odor like cedarwood.

Hygrocybe singeri var. ***singeri*** (A. H. Smith and Hesler) Singer Illus. p. 108

SYNONYMS: *Hygrophorus singeri* A. H. Smith and Hesler

COMMON NAME: Singer's Waxcap.

CAP: 1–3 cm wide, conic, becoming broadly conic; surface glabrous, glutinous to distinctly viscid, translucent-striate to the disc, reddish orange to yellow, blackening in age; flesh very soft, greenish yellow, blackening when exposed; odor and taste not distinctive.

GILLS: strongly adnexed to nearly free, typically attached only at the apex of the stalk, close, broad, pale yellow at first, becoming greenish yellow, staining black where bruised.

STALK: 4–6 cm long, 3–5 mm thick, nearly equal overall, slimy-viscid over the entire length, pale orange-yellow, becoming greenish yellow, staining black where handled or bruised.

MICROSCOPIC FEATURES: spores 9–12 × 5–6 µm, ellipsoid, smooth, hyaline, inamyloid.

OCCURRENCE: solitary or scattered in wet habitats, especially among sphagnum mosses in bogs or wet woodlands; July–October; Michigan and New York, distribution limits and frequency yet to be determined.

EDIBILITY: unknown.

OBSERVATIONS: *Hygrocybe conica* var. *conica* (p. 26) and other species having conic caps, similar colors, and blackening of the fruitbody lack the slimy-viscid stalk. As Hesler and Smith (1963) point out in their description of *Hygrophorus singeri,* the stalk is so slippery that it is difficult to hold. This species is named after the great German mycologist Rolf Singer. The photos show collections from the West Coast, where this waxcap is apparently more often encountered than in the eastern United States.

Hygrocybe spadicea var. ***spadicea*** (Scopoli) P. Karsten Illus. p. 93

SYNONYMS: *Hygrophorus spadiceus* (Scopoli) Fries

COMMON NAME: Date-Colored Waxcap.

CAP: 2.5–7 cm wide, conic, becoming broadly conic to flattened, usually with an umbo; margin often splitting radially; surface viscid then dry, olive brown to grayish brown or fuscous, sometimes with yellow tints; flesh fairly thick, fragile, pale greenish yellow, not staining when exposed or bruised; odor and taste not distinctive.

GILLS: adnexed to nearly free, close, broad, yellow, sometimes becoming pale orange-yellow in age; edges often serrate or eroded.

STALK: 4–8 cm long, 6–10 mm thick, nearly equal overall or slightly tapered in either direction, hollow or stuffed, fragile and easily splitting, colored like the gills or paler, usually overlaid with longitudinal olive brown fibers, with a whitish base.

MICROSCOPIC FEATURES: spores 8–12 × 5–8 µm, ellipsoid to oblong, smooth, hyaline, inamyloid.

OCCURRENCE: scattered or in groups on the ground in burned areas, including broadleaf woodlands and blueberry fields, sometimes associated with staghorn sumac; June–September; reported from Ontario, Maine, Michigan, and Tennessee, distribution limits yet to be determined; rare.

EDIBILITY: unknown.

OBSERVATIONS: *Hygrocybe spadicea* var. *spadicea* f. *odora* (A. H. Smith and Hesler) comb. nov. is nearly identical but has a glabrous stalk and a sharp, radish-like odor. *Hygrocybe spadicea* var. *spadicea* f. *albifolia* (Hesler and A. H. Smith) comb. nov. is also nearly identical but has white gills, flesh, and stalk. Compare with *Hygrophorus hypothejus* var. *hypothejus* (p. 66), which has decurrent gills and a viscid to glutinous stalk and grows in association with conifer trees in late fall or winter. The name *spadicea* means “chestnut brown” or “date brown.”

Hygrocybe squamulosa (Ellis and Everhhart) Arnolds Illus. p. 112

SYNONYMS:

Camarophyllus squamulosus (Ellis and Everhart) Murrill
Hygrophorus miniatus var. *firmus* A. H. Smith and Hesler
Hygrophorus squamulosus Ellis and Everhart

COMMON NAME: none.

CAP: 1.5–5 cm wide, obtuse to convex, sometimes slightly depressed over the disc; margin incurved and remaining so well into maturity; surface dry or slightly moist, glabrous when young but soon developing minute and often distinct fibrillose squamules except on the margin, bright red to orange-red to bright orange-yellow, sometimes with yellow along the margin, fading to yellow overall in age; flesh thick, firm, colored like the cap surface then fading to yellow; odor and taste not distinctive.

GILLS: adnate to adnexed with a decurrent tooth, sometimes free, close to subdistant, broad, reddish to pale yellow; edges even.

STALK: 3–5 cm long, 3–6 mm thick, nearly equal overall, often compressed, hollow, white-pruinose near the apex and glabrous elsewhere, apricot yellow or tinged with red, base sometimes coated with a white tomentum.

MICROSCOPIC FEATURES: spores 6–9 × 4–5 µm, subellipsoid, primarily binucleate, smooth, hyaline, inamyloid.

OCCURRENCE: scattered or in groups on the ground or sometimes on decaying logs in woodland or among sphagnum mosses in bogs; July–October; widely distributed in the Northeast; occasional.

EDIBILITY: unknown.

OBSERVATIONS: *Hygrocybe miniata* var. *miniata* (p. 36) is similar but has a thinner, smooth to slightly scurfy cap and uninucleate spores that are often constricted near the middle. *Hygrocybe cantharellus* f. *cantharellus* (p. 20) is also similar but has less conspicuous squamules, strongly decurrent gills, and uninucleate spores. The name *squamulosa* means "full of scales," in reference to the scaly cap.

Hygrocybe subaustralis (A. H. Smith and Hesler) comb. nov. Not Illustrated

SYNONYMS: *Hygrophorus subaustralis* A. H. Smith and Hesler

COMMON NAME: none.

CAP: 1–5 cm wide, conic, then convex to nearly plane; surface matted-fibrillose, shiny, at times finely powdered, dry or moist, not viscid, white with a yellowish center; flesh white, waxy; odor not distinctive; taste bitter.

GILLS: adnexed to emarginate, close to subdistant, thin, broad, waxy, white becoming cream-colored; edges even.

STALK: 2–10 cm long, 2–6 mm thick, equal, stuffed, becoming hollow, appressed-fibrillose, apex pruinose, dry, shiny, white.

MICROSCOPIC FEATURES: spores 5–6.5 × 4.5–6 µm, ellipsoid to subovoid, smooth, hyaline, inamyloid.

OCCURRENCE: scattered on bare soil in broadleaf and mixed woods; August–September; North Carolina and Tennessee; locally common.

EDIBILITY: unknown.

OBSERVATIONS: The adnexed to emarginate gill attachment helps to distinguish this species from other white waxcaps. The name *subaustralis* means "nearly southern," in reference to this small waxcap's known distribution.

Hygrocybe subovina (Hesler and A. H. Smith) Lodge and Cantrell

Illus. p. 95

SYNONYMS: *Hygrophorus subovinus* Hesler and A. H. Smith
COMMON NAME: Brown Sugar Waxcap.
CAP: 2–5 cm wide, convex to broadly convex to flat in age; surface dry, fibrillose, squamulose over the disc, appearing smooth at times, dark gray brown when dry to nearly black when moist; flesh thin at the margin, thick at the center, brittle, unchanging when bruised, concolorous with the cap or paler; odor of brown sugar; taste slightly alkaline.
GILLS: adnexed, becoming notched, distant, broad, thick, edges even, whitish to grayish brown, bruising reddish.
STALK: 3–6 cm long, 4–9 mm thick, dry, equal, hollow, cylindrical or compressed at times, concolorous with the cap or darker, especially toward the base, not staining when handled or bruised.
MICROSCOPIC FEATURES: spores 5–6 × 5–5.7 µm, globose to subglobose, smooth, hyaline, inamyloid.
OCCURRENCE: scattered or in groups on the ground in broadleaf woodlands, often with oak; June–October; widely distributed but more common in the southern range; occasional to locally common.
EDIBILITY: unknown.
OBSERVATIONS: The sweet odor helps distinguish this easily overlooked dark-colored waxcap from *Hygrocybe ovina* (p. 42), which is variously described as having a fruity, ammoniac, or sheep-like odor. Compare with *Hygrocybe nitrata* (p. 41), which smells of ammonia. The name *subovina* refers to the similarity with *Hygrocybe ovina*.

Hygrocybe tahquamenonensis (A. H. Smith and Hesler) comb. nov.

Illus. p. 103

SYNONYMS: *Hygrophorus tahquamenonensis* A. H. Smith and Hesler
COMMON NAME: none.
CAP: 2–6 cm wide, obtusely conic; surface moist and hygrophanous, becoming squamulose when faded, pale ocher yellow and fading to pinkish buff; margin opaque and flaring, edge becoming black when bruised; flesh thin, yellowish, becoming dingy pallid; odor and taste nitrous.
GILLS: adnate to adnexed, broad, distant, ventricose, pale yellow fading to pale pinkish buff; edges entire.

STALK: 4–5 cm long, 4–8 mm thick, nearly equal or slightly enlarged downward, becoming hollow, glabrous, pallid to pallid yellowish, staining brownish where handled.

MICROSCOPIC FEATURES: spores 7–9 × 5–5.5 µm, oblong to subovoid, smooth, hyaline, inamyloid; clamp connections present.

OCCURRENCE: single or scattered in mixed woods; August–September; reported from Michigan and Quebec, distribution limits unknown.

EDIBILITY: unknown.

OBSERVATIONS: *Hygrocybe nitrata* (p. 41) is similar except for its gray to brown colors and lack of dark staining on the cap margin and stalk. The name *tahquamenonensis* refers to Tahquamenon Falls State Park in Michigan, where this waxcap was first collected.

Hygrocybe virginea var. **_virginea_** (Wülfen : Fries) P. D. Orton and Watling

Illus. p. 83

SYNONYMS:

Camarophyllus borealis (Peck) Murrill
Camarophyllus niveus (Fries) Wünsche
Hygrocybe borealis f. *borealis* (Peck) Bon
Hygrophorus borealis Peck
Hygrophorus niveus Fries
Hygrophorus virgineus (Wülfen : Fries) Fries

COMMON NAME: Snowy Waxcap.

CAP: 1–7 cm wide, hemispherical when young, becoming broadly convex to nearly plane, sometimes with a depressed center or small umbo; margin incurved at first and often remaining so well into maturity, sometimes uplifted in age, translucent-striate halfway to the disc when fresh; surface moist to lubricous, hygrophanous, white or cream to ivory, occasionally with pale brownish tinges, becoming white when dry; flesh thick or thin, white; odor and taste not distinctive.

GILLS: decurrent, subdistant, sometimes forked, white to whitish; edges even.

STALK: 2–6 cm long, 2–10 mm thick, nearly equal or slightly tapered in either direction, sometimes compressed, dry, glabrous, white to whitish.

MICROSCOPIC FEATURES: spores 6–9 × 4–6 µm, ellipsoid to ellipsoid-oblong, smooth, hyaline, inamyloid.

OCCURRENCE: scattered or in groups in grassy areas, among mosses, or on the ground in conifer or broadleaf woods; July–December; widespread; occasional.

EDIBILITY: edible.

OBSERVATIONS: This mushroom is highly variable in stature, thickness of the flesh, and shades of color. In the past, larger and stouter specimens have been called *Hygrophorus virgineus*, whereas smaller and more slender specimens have been called *H. niveus. Hygrocybe virginea* var. *fuscescens* (Bresadola) Arnolds is similar but tends to be smaller and more fragile. It has a grayish brown to cinnamon brown disc with brownish radial striations and a stalk that is often tapered downward. *Hygrophorus eburneus* var. *eburneus* (p. 131) is also similar, but it has a pure white, viscid to glutinous cap with an inrolled to incurved margin and a viscid to glutinous pure white stalk. *Hygrocybe borealis* f. *borealis* (Peck) : Bon = *Hygrophorus borealis* Peck has been commonly misidentified for many years in North America. Boertmann (1995) has placed *Hygrocybe nivea (Hygrophorus niveus)* in synonymy with *H. virginea* var. *virginea,* a move with which we agree. Singer (1951) and others have questioned whether *Hygrocybe borealis* f. *borealis* is different from *H. nivea*. We believe that *Hygrocybe borealis* f. *borealis* should also be placed in synonymy with *H. virginea* var. *virginea*. We cannot find a single good character to separate them. Hesler and A. H. Smith (1963) state that *Hygrocybe borealis* f. *borealis* is related to *H. virginea* var. *virginea* but differs by having a more slender stalk and smaller spores. After carefully comparing the stalk diameters and spore sizes of these two species, we are of the opinion that they are essentially identical. Therefore, we have placed *Hygrocybe borealis* f. *borealis* in synonymy with *H. virginea* var. *virginea*. We have included a photo of *Hygrocybe virginea* f. *salmonea* (Coker) comb. nov. (p. 84), which was originally described under the name *Hygrophorus borealis* f. *salmoneus*. The name *virginea* means "purest white."

HYGROPHORUS

Hygrophorus agathosmus Fries — Illus. p. 140

SYNONYMS: *Hygrophorus cerasinus* (Berkeley) Fries

COMMON NAME: Gray Almond Waxcap.

CAP: 2–8 cm wide, convex, becoming broadly convex to nearly flat, at times with a slight umbo; surface smooth, glutinous to viscid, ashy gray to brownish gray or grayish tan, darker on the disc and fading toward the margin; margin inrolled and slightly velvety when young; flesh whitish gray to white, soft; odor of almond extract or cherry pits; taste not distinctive.

GILLS: adnate to slightly decurrent, subdistant, narrow, sometimes forked, waxy, creamy white.

STALK: 5–8 cm long, 5–20 mm thick, nearly equal, solid, dry, fibrillose to pruinose when young, smooth in age except at the apex, whitish to ashy gray.

MICROSCOPIC FEATURES: spores 7–10 × 4.5–5.5 µm, ellipsoid, hyaline, inamyloid.

OCCURRENCE: solitary, scattered, or in groups on the ground under conifers, especially spruce and pine; July–December; widely distributed; occasional.

EDIBILITY: reported to be edible.

OBSERVATIONS: *Hygrophorus bakerensis* (p. 56), which also has an odor of almond extract or cherry pits, differs in having cap colors that are various shades of brown, not gray. *Hygrophorus monticola* (p. 69) is also similar and has an odor of almond extract or cherry pits, but its cap is cinnamon buff on the disc and is flushed with vinaceous red or salmon tones over a brownish ground color that extends to the margin in age, and it has larger spores that measure 10–14 × 5.5–7.5 µm. The name *agathosmus* means "having an agreeable odor."

Hygrophorus albofuscus Hesler and A. H. Smith Illus. p. 139

SYNONYMS: *Hygrophorus fuscoalbus* var. *occidentalis* Kauffman

COMMON NAME: none.

CAP: 2–5 cm wide, convex, becoming broadly convex to nearly plane, sometimes slightly depressed over the disc; margin at first involute and floccose-downy, even; surface viscid, glabrous, grayish brown to brownish ashy, sometimes blackish on the disc; flesh thin, white; odor and taste not distinctive.

GILLS: adnate to decurrent, subdistant to close, rather narrow, intervenose, creamy white.

STALK: 3–7 cm long, 4–12 mm thick, slender or sometimes stout, nearly equal or tapered downward, dry, solid, sometimes curved at the base, somewhat fragile, glabrous except for the apex, which is floccose-pruinose, white or whitish.

MICROSCOPIC FEATURES: spores 5.5–8 × 3–4.5 µm, ellipsoid, smooth, hyaline, inamyloid.

OCCURRENCE: scattered or in groups on the ground in broadleaf woods, especially with oak; October; Michigan, distribution limits and frequency yet to be determined.

EDIBILITY: unknown.

OBSERVATIONS: *Hygrophorus fuscoalbus* var. *fuscoalbus* (Lasch) Fries (p. 64) is similar but has a dark olive brown to grayish brown cap with a glutinous sheath on its stalk and much larger spores that measure 9–13 × 5.5–7 µm. *Hygrophorus laurae* (p. 68) is also similar but has a white to whitish cap with a brownish disc and a viscid to glutinous stalk. The term *albofuscus* means "whitish brown."

Hygrophorus bakerensis A. H. Smith and Hesler Illus. p. 147

SYNONYMS: none.

COMMON NAME: Tawny Almond Waxcap.

CAP: 4–15 cm wide, rounded when young, becoming nearly plane in age; surface slimy when moist, covered with small, flattened, brownish fibers beneath the gluten, yellow-brown or cinnamon brown or shades of orange-brown on the disc, becoming whitish to cream toward the margin; margin incurved and cottony when young, becoming smooth and flat in age; flesh thick, firm, white; odor strongly fragrant, resembling almond extract or cherry pits; taste not distinctive.

GILLS: decurrent, close to subdistant, waxy, whitish to cream.

STALK: 4–14 cm long, 8–25 mm thick, equal or tapering downward, solid, dry, cottony at the apex when young, smooth overall in age, often with clear drops of liquid when moist, white to cream overall.

MICROSCOPIC FEATURES: spores 7–10 × 4.5–6 µm, ellipsoid, smooth, hyaline, inamyloid.

OCCURRENCE: scattered or in groups on the ground under conifers; August–December; widely distributed; occasional.

EDIBILITY: reported to be edible.

OBSERVATIONS: Compare with *Hygrophorus monticola* (p. 69), which is similar and has an odor of almond extract or cherry pits, but its cap is cinnamon buff on the disc and is flushed with vinaceous red or salmon tones that are over a brownish ground color and that extend to the margin in age, and it has larger spores that measure 10–14 × 5.5–7.5 µm. *Hygrophorus agathosmus* (p. 55) is also similar. It has an odor of almond extract or cherry pits but has an ashy gray to brownish gray or grayish tan cap. Compare also with *Hygrophorus tennesseensis* (p. 80).

Hygrophorus camarophyllus (Albertini and Schweinitz) Dumée Illus. p. 149

SYNONYMS: *Hygrophorus burnhami* Peck

COMMON NAME: Dusky Waxcap.

CAP: 4–13 cm wide; convex to broadly convex, occasionally with an umbo; surface dry, viscid when moist, smooth but appearing streaked and minutely downy, tan to brownish gray to coffee brown or vinaceous brown; margin downy at first; flesh thick, white; odor of coal tar or not distinctive; taste somewhat sweetish or not distinctive.

GILLS: adnate becoming subdecurrent, close to subdistant, broad, waxy, white to cream or grayish.

STALK: 3–13 cm long, 1–2 cm thick, equal or tapered downward, solid, dry, colored like the cap or paler.

MICROSCOPIC FEATURES: spores 7–9 × 4–5 µm, ellipsoid, smooth, hyaline, inamyloid.

OCCURRENCE: scattered or in groups on the ground under conifers; July–December; widely distributed; occasional.

EDIBILITY: edible.

OBSERVATIONS: A darkly colored cap, white to cream or grayish gills, and odor of coal tar (when present) are distinctive features. The name *camarophyllus* means "arched leaves," in reference to mature specimens' more or less decurrent gills.

Hygrophorus capreolarius (Kalchbrenner) Saccardo Illus. p. 158

SYNONYMS: *Hygrophorus purpureobadius* S. Imai

COMMON NAME: none.

CAP: 3–7 cm wide, convex with an inrolled margin, becoming nearly plane, broadly umbonate at times; surface viscid when young and fresh, becoming dry in age, color variable, from pink to wine red with purple fibers; flesh thick, whitish to concolorous with the cap surface, spongy; odor and taste not distinctive.

GILLS: adnate to decurrent, subdistant, may be intervenose, narrow to moderately broad, fairly thick, brittle, pinkish, becoming concolorous with the cap; edges entire.

STALK: 3.5–10 cm long, 7–14 mm thick, nearly equal or tapered downward and sometimes subradicating, dry, fibrillose, lacking a partial veil or ring, at first whitish with a reddish tinge, becoming concolorous with the cap at maturity, quickly yellow with the application of KOH.

MICROSCOPIC FEATURES: spores 6–9 × 4–5 µm, ellipsoid, smooth, hyaline, inamyloid.

OCCURRENCE: gregarious among mosses or on the ground, usually under spruce; September–December; widely distributed across northeastern North America, also reported from Europe and Asia; uncommon.

EDIBILITY: reported to be edible.

OBSERVATIONS: *Hygrophorus purpurascens* (p. 73) is similar and occurs in similar habitats, but it has a partial veil and overall lighter coloration. Compare with *Hygrophorus erubescens* var. *erubescens* (p. 61), which is paler and usually stains yellow on the stalk or flesh when bruised. Compare also with *Hygrophorus russula* (p. 76), which grows in association with broadleaf trees. The name *capreolarius* refers to the smell of goats, but the application here is unclear.

Hygrophorus chrysaspis Métrod Illus. p. 134

SYNONYMS: none.

COMMON NAME: none.

CAP: 2–9 cm wide, convex to nearly plane with an inrolled margin at first; surface glutinous to viscid, white becoming pale pinkish buff, center possibly drying to dark gray to blackish in age or streaked black; flesh watery cream, staining yellow at times, staining yellow to brownish orange with the application of KOH; odor faintly fragrant or not distinctive; taste not distinctive.

GILLS: adnate, subdistant to close, medium broad, white, becoming yellow, bruising yellow, dark reddish brown when dried; edges entire.

STALK: 5–10 cm long, 2–8 mm thick, nearly equal or tapered downward, solid, becoming hollow in age, glutinous, glabrous, apex pruinose and beaded with pale yellow droplets, white, becoming pinkish buff to yellow when bruised.

MICROSCOPIC FEATURES: spores 7–9 × 3.5–4.5 µm, ellipsoid, smooth, hyaline, inamyloid.

OCCURRENCE: on soil under beech and other broadleaf trees; August–October; Michigan and Tennessee; locally common.

EDIBILITY: unknown.

OBSERVATIONS: Compare with *Hygrophorus eburneus* var. *eburneus* (p. 61), which is similar, but its cap and gills are white overall and become only slightly yellowish on drying. The name *chrysapsis* derives from the Greek word *chryso,* meaning "gold," probably referring to the gills, which become yellow with age.

Hygrophorus chrysodon (Batsch) Fries Illus. p. 134

SYNONYMS: *Hygrophorus chrysodon* var. *leucodon* Albertini and Schweinitz

COMMON NAME: Golden-Spotted Waxcap.

CAP: 3–8 cm wide; convex, becoming broadly convex to nearly flat; surface viscid when moist, shiny when dry, smooth, white, covered with yellow-orange to golden yellow granules over the entire surface when young, which often remain on the margin and on the disc in age; margin inrolled and floccose-tomentose initially, expanding in age; flesh white, thick; odor and taste not distinctive.

GILLS: decurrent, distant, moderately broad, white to cream, edges sometimes fringed with yellow.

STALK: 3–8 cm long, 5–15 mm thick, equal, stuffed, viscid or dry, white with golden yellow granules at the apex, sometimes dark yellow at the base.

MICROSCOPIC FEATURES: spores 7–11 × 3.5–6 µm, ellipsoid, smooth, hyaline, inamyloid.

OCCURRENCE: solitary, scattered, or in groups on the ground under broadleaf and conifer trees; July–December; widely distributed; fairly common.

EDIBILITY: reported to be edible.

OBSERVATIONS: Some authors have recognized rare collections with white instead of yellow granules on the cap margin as *Hygrophorus chrysodon* var. *leucodon*. We do not believe that this character alone merits a distinct varietal designation and have placed *Hygrophorus chrysodon* var. *leucodon* in synonymy. The term *chrysodon* means "gold tooth," referring to the prominent yellow granules on the cap margin.

Hygrophorus cossus Fries Illus. p. 131

SYNONYMS: *Hygrophorus eburneus* var. *cossus* (Sowerby ex. Fries) Quélet

COMMON NAME: Goat Moth Waxcap.

CAP: 3–7 cm wide, convex to nearly plane; margin inrolled at first, becoming decurved in age; surface glutinous, glabrous white, buff at the center; flesh white, soft, unchanging when bruised; odor strong and unpleasant, variously described as anything from aromatic to like shellfish or Jerusalem artichoke; taste not distinctive.

GILLS: adnate to decurrent, close to subdistant, narrow to broad, white or cream-colored, becoming tinted salmon buff; edges even.

STALK: 4–9 cm long, 8–12 mm thick, solid, nearly equal or tapered downward, punctate to scabrous at the apex, lower two-thirds with a gelatinous sheath that covers a thin white fibrillose veil, white to yellowish, base salmon buff to brownish in age.

MICROSCOPIC FEATURES: spores 7–10 × 4–5 µm, ellipsoid, smooth, hyaline, inamyloid.

OCCURRENCE: scattered to gregarious under conifers or in mixed woods; September–October; Michigan, also reported from Idaho and California; rare in eastern North America.

EDIBILITY: not edible.

OBSERVATIONS: Some mycologists consider this waxcap an odorous variety of *Hygrophorus eburneus.* The common and scientific names refer to the similarity of its odor to that of a Goat Moth caterpillar *(Cossus cossus),* which occurs in Europe, North Africa, and Asia.

Hygrophorus cremicolor (Murrill) Murrill Not Illustrated

SYNONYMS: *Camarophyllus cremicolor* Murrill

COMMON NAME: none.

CAP: 1–3 cm wide, convex, becoming umbonate; margin inrolled at first, floccose; surface canescent, moist, hygrophanous, white to cream, sometimes tinted salmon; flesh cream to pinkish buff; odor and taste not distinctive.

GILLS: decurrent, distant, narrow, yellow; edges even.

STALK: 5–7 cm long, 8–12 mm thick, nearly equal or tapered downward, glabrous, concolorous with the cap, sometimes paler toward the base.

MICROSCOPIC FEATURES: spores 5–7 × 3.5–4.5 µm, ellipsoid to subglobose, smooth, hyaline, inamyloid.

OCCURRENCE: on soil in woods and wet areas; July–November; Massachusetts and Michigan; uncommon.

OBSERVATIONS: This small waxcap resembles a pale *Hygrocybe pratensis* var. *pratensis* (p. 43), which is a much more common species but lacks the floccose margin. The name *cremicolor* refers to the color of cream.

Hygrophorus discoideus (Persoon) Fries Illus. p. 146

SYNONYMS: none.

COMMON NAME: Clay Waxcap.

CAP: 2–5 cm wide, convex to nearly plane, with a slight umbo at times; margin inrolled and becoming elevated at maturity; surface glutinous to viscid, center glabrous, margin pubescent, reddish brown to brown at the center, buff to yellowish pink or pale orange-yellow at the margin; flesh buff or tinged orange to reddish brown, thin; odor and taste not distinctive.

GILLS: adnate to subdecurrent, close to subdistant, narrow to fairly broad, waxy, white or tinged yellowish pink or buff; edges even.

STALK: 4–9 cm long, 3–8 mm thick, nearly equal overall, solid, becoming hollow at maturity, surface fibrillose and dry at the apex, viscid below from the remnants of the evanescent glutinous veil, white to pale ochraceous, at times staining pale yellow at the base.

MICROSCOPIC FEATURES: spores 5.5–8 × 3.3–5 µm, ellipsoid, smooth, hyaline, inamyloid.

OCCURRENCE: scattered to gregarious in sphagnum mosses and in humus under spruce and other conifers; September–November; Michigan and Ontario; locally common.

EDIBILITY: unknown.

OBSERVATIONS: Compare with *Hygrophorus hypothejus* var. *hypothejus* (p. 66), which has larger spores, is usually associated with two-needle pines, and fruits late in the season. *Hygrophorus discoideus* also bears a superficial resemblance to *Hebeloma mesophaeum* (Persoon) Quélet, which has brown spores. The word *discoideus* means "disk-like."

Hygrophorus eburneus* var. *eburneus (Bulliard) Fries Illus. p. 131

SYNONYMS: none

COMMON NAME: Ivory Waxcap, White Waxcap.

CAP: 2–7 cm wide, obtuse to convex, becoming broadly convex to nearly plane, sometimes with a low umbo; margin even, inrolled to incurved and floccose-pubescent when young; surface glutinous to viscid, shiny, smooth, pure white overall or sometimes duller over the disc; flesh white, unchanging when exposed; odor and taste not distinctive.

GILLS: decurrent, subdistant to distant, white when young, buff to yellowish in age.

STALK: 4.5–11.5 cm long, 4–15 mm thick, nearly equal or tapered downward, sometimes abruptly narrowed at the base, glutinous, silky beneath the gluten, hollow in age, pure white.

MICROSCOPIC FEATURES: spores 6–9 × 3.5–5 µm, ellipsoid, smooth, hyaline, inamyloid.

OCCURRENCE: scattered or in groups on the ground in conifer and broadleaf forests or mixed woodlands; August–December; widely distributed; occasional.

EDIBILITY: edible.

OBSERVATIONS: *Hygrocybe virginea* var. *virginea* (p. 54) has a moist to dry but not viscid whitish cap that often has pale yellow to pale tan on the disc and a translucent-striate margin when moist. *Hygrophorus chrysaspis* (p. 58) has watery cream flesh that stains yellow to brownish orange with the application of KOH and white gills that become dark reddish brown when dried. *Hygrophorus cossus* (p. 60) is similar but has a pronounced sweetish or unpleasant odor. Compare with *Hygrophorus piceae* (p. 71), which grows with spruce and has a dry or only slightly viscid stalk. The name *eburneus* means "ivory white."

Hygrophorus erubescens var. ***erubescens*** (Fries) Fries Illus. p. 159

SYNONYMS: none.

COMMON NAME: none.

CAP: 2–9 cm wide, convex, becoming broadly convex with a low broad umbo; surface glutinous to viscid, color variable, spotted and streaked vinaceous pink to purplish red over a pale pink to white ground color, typically white along the margin and darkest on the disc; margin incurved and white when young; flesh white, usually staining yellow when cut and rubbed; odor and taste not distinctive.

GILLS: adnate at first, becoming decurrent, close to subdistant, medium broad, whitish to pinkish buff when young, becoming spotted pinkish red in age.

STALK: 4–9 cm long, 1–2 cm thick, nearly equal down to a tapered base, dry, scurfy overall or at least on the upper portion, white and beaded with moisture drops at the apex when fresh, white below the apex with pinkish red fibrils and dots, sometimes yellow, especially at the base, in age or when bruised.

MICROSCOPIC FEATURES: spores 7–11 × 5–6 µm, ellipsoid, smooth, hyaline, inamyloid.

OCCURRENCE: scattered or in groups on the ground under conifers, especially pine and spruce; July–November; widely distributed in the Northeast; occasional.

EDIBILITY: unknown.

OBSERVATIONS: *Hygrophorus erubescens* var. *gracilis* (p. 63) has a much narrower cap, a thinner and typically longer stalk, distant gills, and flesh that does not stain yellowish when injured. *Hygrophorus russula* (p. 76) is similar but has close to crowded gills, grows on the ground under broadleaf trees, especially oaks, has a more compact appearance with a larger cap and thicker stalk, and flesh that does not stain yellow when exposed or bruised. Compare with *Hygrophorus purpurascens* (p. 73), which has a fleeting partial veil that is conspicuous only on very young specimens. The name *erubescens* means "blushing."

Hygrophorus erubescens var. ***gracilis*** A. H. Smith and Hesler Not Illustrated

SYNONYMS: none.

COMMON NAME: none.

CAP: 2–3 cm wide, broadly convex to nearly plane, sometimes slightly depressed or umbonate; surface viscid, dark vinaceous pink to purplish red over the disc and paler vinaceous pink outward; margin whitish at first then colored like the disc, decurved and typically remaining so well into maturity; flesh white, not staining when injured; odor and taste not distinctive.

GILLS: adnate at first then becoming decurrent, distant, narrow, white or flushed grayish pink, staining reddish when bruised or in age.

STALK: 6–10 cm long, 4–8 mm thick, nearly equal or tapered downward, typically with a narrowed and pointed base, dry, whitish then becoming purplish red in age; upper portion with white fibrillose scales or coarsely fibrillose-pruinose; lower portion appressed-fibrillose.

MICROSCOPIC FEATURES: spores 8–11 × 5.5–7 µm, ellipsoid, smooth, hyaline, inamyloid.

EDIBILITY: unknown.

OCCURRENCE: scattered or in groups on the ground under pines; October–December; reported from Tennessee and the Pacific Northwest, but distribution limits and frequency yet to be determined.

OBSERVATIONS: *Hygrophorus erubescens* var. *erubescens* (p. 62) has a much wider cap, a thicker and typically much shorter stalk, close to subdistant gills, and flesh that usually stains yellow when injured. The name *gracilis* means "slender" or "graceful," in reference to this waxcap's stature.

Hygrophorus flavodiscus Frost Illus. p. 135

SYNONYMS: none.

COMMON NAME: Yellow-Centered Waxcap.

CAP: 2–7 cm wide, convex at first, becoming broadly convex to nearly plane at maturity; surface glutinous, white to whitish with a pale yellow or reddish yellow disc that becomes ochraceous orange to orange-buff on drying, sometimes with grayish streaks when the hyaline gluten dries; flesh thick, firm, white; odor and taste not distinctive.

GILLS: adnate to decurrent, subdistant, broad, covered by a layer of gluten when specimens are very young, pale pinkish at first, soon white to whitish.

STALK: 3–7.5 cm long, 6–15 mm thick, nearly equal overall or sometimes tapered downward, sheathed nearly to the apex with a layer of gluten, fibrillose to scurfy beneath the gluten, white to whitish and sometimes with pale brown to dull yellow tinges on older specimens.

MICROSCOPIC FEATURES: spores 6–8 × 3.5–5 µm, ellipsoid, smooth, hyaline, inamyloid.

OCCURRENCE: scattered or in groups on the ground or among mosses in conifer woods, especially with white pine; September–November; widely distributed in the Northeast; fairly common.

EDIBILITY: edible.

OBSERVATIONS: This mushroom fruits late in the season, often during cold weather and sometimes after frosts. *Hygrophorus gliocyclus* (p. 65) is similar but has a pale brownish yellow to dull golden yellow or ochraceous yellow cap with a paler margin, gills that are ivory yellow to dingy yellow, and larger spores that measure 8–11 × 4.5–6 µm. The name *flavodiscus* means "yellowish disc."

Hygrophorus fuligineus Frost Illus. p. 149

SYNONYMS: none.

COMMON NAME: Sooty Waxcap.

CAP: 4–13 cm wide, orbicular or convex to nearly plane; surface glabrous, covered with a thick hyaline gluten, blackish brown at first, becoming olivaceous brown, margin paler; flesh thick, white, tinted gray near the cap surface; odor and taste not distinctive.

GILLS: adnate to subdecurrent, subdistant, attenuated, covered by gluten in the button stage, white to cream-colored.

STALK: 4–9 cm long, 1–2.5 cm thick, nearly equal, solid, surface glutinous below the apex from the slime veil, white to pale brownish, apex dry and silky to scabrous with an obscure annular zone where the slime ends.

MICROSCOPIC FEATURES: spores 7–9 × 4.5–5.5 µm, short ellipsoid, smooth, hyaline, inamyloid.

OCCURRENCE: scattered and often gregarious under conifers, especially white pine *(Pinus strobus)*; August–January; widely distributed in eastern North America; common.

EDIBILITY: edible, but the heavy slime layer will deter many from trying it.

OBSERVATIONS: This late-season waxcap often fruits with *Hygrophorus flavodiscus* (p. 63). In wet conditions, it is viscid and difficult to collect. It usually fruits in cool-weather conditions late in the season. Buttons resemble shiny black marbles. *Hygrophorus paludosus* (p. 71) is similar, but its glutinous cap is pale ochraceous buff to pinkish vinaceous and appears to be streaked or netted beneath the gluten. The term *fuligineus* means "sooty" or "blackish."

Hygrophorus fuscoalbus var. ***fuscoalbus*** (Lasch) Fries Illus. p. 153

SYNONYMS: *Hygrophorus latitabundus* Britzelmayr

COMMON NAME: none.

CAP: 2.5–5 cm wide, convex to nearly plane; surface viscid, glabrous except floccose at the margin, dark olive brown to grayish brown, darkest over the disc and paler toward the margin, becoming ashy gray with blackish tinges in age; flesh thick at the center, grayish white; odor faintly aromatic or not distinctive; taste not distinctive.

GILLS: adnate to decurrent, close, broad, thick, white, tinged pinkish buff at times.

STALK: 3–6 cm long, 8–15 mm wide, usually nearly equal overall but sometimes tapered in either direction, compressed at times, covered with a glutinous sheath up to an apical annular zone of whitish fibrils, solid, white.

MICROSCOPIC FEATURES: spores 8–13 × 5–7 µm, ellipsoid, smooth, hyaline, inamyloid

OCCURRENCE: scattered on soil in conifer and mixed woods, often under pine; October–November; reported from New York, Ontario, Michigan, western North America, and Europe; uncommon.

EDIBILITY: reported as edible in Europe.

OBSERVATIONS: There are taxonomic issues with this species. North American collections may differ from *Hygrophorus latitabundus* Britzelmayr, which Eef Arnolds considers to be a synonym. *Hygrophorus albofuscus* (p. 56) is similar but has a dry stalk that lacks a glutinous sheath and much smaller spores that measure 5.5–8 × 3–4.5 µm. The name *fuscoalbus* means "grayish brown" and "white."

Hygrophorus gliocyclus Fries

Illus. p. 136

SYNONYMS: none.

COMMON NAME: none.

CAP: 2–9 cm wide, convex, becoming nearly flat, sometimes with a very broad umbo; margin involute at first, even; surface smooth, glutinous, pale brownish yellow to dull golden yellow or ochraceous yellow, paler near the margin; flesh white; odor and taste not distinctive.

GILLS: decurrent, subdistant, broad, ivory yellow to more dingy yellowish as specimens age.

STALK: 2–6 cm long, 6–12 mm thick, nearly equal or somewhat tapered downward, narrowing sharply at the base, solid, whitish or sometimes colored like the cap, covered nearly to the apex with a hyaline glutinous layer.

MICROSCOPIC FEATURES: spores 8–11 × 4.5–6 µm, ellipsoid, smooth, hyaline, inamyloid.

OCCURRENCE: in groups or clusters on the ground under spruce and pine; August–December; widely distributed; occasional.

EDIBILITY: edible.

OBSERVATIONS: *Hygrophorus flavodiscus* (p. 63) is similar but has a white cap with a pale yellow or reddish yellow disc, gills that are pinkish when young then become whitish, and smaller spores that measure 6–8 × 3.5–5 µm. The name *gliocyclus* means "glue wheel," relating to the glutinous cap.

Hygrophorus glutinosus Peck Illus. p. 136

SYNONYMS: *Hygrophorus rubropunctus* Peck

COMMON NAME: none.

CAP: 4–5 cm wide, convex, broadly umbonate at times; margin involute at first, even; surface glabrous, glutinous, creamy buff at the center, yellowish cream outward, bright yellow when dry; flesh thick over the center, thin toward the margin, white; odor and taste not distinctive.

GILLS: adnate or adnexed, close, becoming subdistant at maturity, white, becoming yellowish in age; edges even.

STALK: 4–9 cm long, 8–15 mm thick, solid, nearly equal or narrowing at the apex, floccose to tomentose, glutinous from the base up nearly to the apex, white with yellow-brown stains, having hyaline drops above the collar that on drying form reddish glandular dots.

MICROSCOPIC FEATURES: spores 8–11 × 5–7 µm, ellipsoid, smooth, hyaline, inamyloid.

OCCURRENCE: solitary to scattered in broadleaf and mixed woods; September–November; New York, North Carolina, and Michigan; uncommon.

EDIBILITY: unknown.

OBSERVATIONS: Compare with *Hygrophorus flavodiscus* (p. 63), which has a white to whitish cap with a pale yellow to reddish yellow disc and a white stalk that lacks the glandular dots at the apex and occurs under pine. The term *glutinosus* means "sticky" or "glutinous."

Hygrophorus hypothejus var. ***hypothejus*** (Fries) Fries Illus. p. 148

SYNONYMS: *Hygrophorus subpustulatus* (Murrill) Murrill

COMMON NAME: Late Fall Waxcap, Herald of Winter.

CAP: 2–6 cm wide, convex, becoming nearly plane or somewhat depressed at maturity, sometimes with a small umbo; surface glutinous when young and fresh, becoming viscid and finally dry in age, olive brown to red-brown on young specimens, fading to pale olive brown to yellow-brown, dull yellow along the margin on mature specimens and remaining brown on the disc; margin sterile and incurved at first, becoming decurved and remaining so well into maturity, often wavy in age; flesh white, unchanging; odor and taste not distinctive.

GILLS: decurrent, subdistant, white at first, soon pale yellow to pale dull yellow, becoming deeper and more intense with age or after exposed to frost, sometimes developing orange stains in age, typically venose on the faces and sometimes intervenose, not staining when cut or bruised.

STALK: 3–5.5 cm long, 7–11 mm thick, nearly equal or tapered slightly in either direction, weakly fibrillose and dry at the apex, covered with a glutinous sheath below the apex when fresh, whitish at first, soon yellowish to pale ochraceous orange on the upper portion and remaining white toward the base.

MICROSCOPIC FEATURES: spores 7–10 × 4–6 µm, ellipsoid, smooth, hyaline, inamyloid.

OCCURRENCE: scattered or in groups under conifers, often in sandy soil or among mosses; August–January; widely distributed; common.

EDIBILITY: edible, but little appreciated.

OBSERVATIONS: In the Southeast, *Hygrophorus hypothejus* var. *hypothejus* is one of the most common *Hygrophorus* species found under two-needle pines during the fall. Compare with *Hygrophorus discoideus* (p. 61). The name *hypothejus* means "sulfur underneath," relating to the yellow tones that are often present on the gills and stalk.

Hygrophorus inocybiformis A. H. Smith Illus. p. 153

SYNONYMS: none.

COMMON NAME: Inocybe-like Waxcap.

CAP: 2–6 cm wide, conic to obtuse at first, becoming broadly convex to campanulate or nearly plane at maturity, broadly umbonate; margin incurved and fringed with remnants of a fibrillose veil; surface appressed-fibrillose to appressed-scaly, dry, dark gray to gray-brown; flesh soft, fragile, thin except in the center area, white to pale gray; odor and taste not distinctive.

GILLS: subdecurrent, subdistant to distant, thick, firm, medium to broad, intervenose at times, white at first, becoming pale grayish in age; edges even.

STALK: 3–6.5 cm long, 5–15 mm thick, nearly equal, solid, dry, apex silky white, lower area streaked with dark gray-brown fibrils over a whitish ground layer.

MICROSCOPIC FEATURES: spores 9.5–17 × 6–9 µm, ellipsoid to oblong with a blunt apiculus, smooth, hyaline, inamyloid.

OCCURRENCE: solitary to scattered or in small groups on the ground or among mosses under hemlock, balsam fir, and spruce; July–November; Nova Scotia, Quebec, Idaho, and California; rarely reported in eastern North America.

EDIBILITY: unknown.

OBSERVATIONS: Although this waxcap is more frequently encountered in western North America, it may be more common than has been reported in the northern conifer forests of eastern North America. *Hygrophorus olivaceoalbus* var. *olivaceoalbus* (p. 70) is similar, but it has a conspicuously streaked cap with smoke gray to black fibrils beneath a layer of gluten, and its stalk has a double sheath consisting of a glutinous outer layer and an inner layer of appressed blackish fibrils. Compare with *Hygrophorus pustulatus* (p. 74). The word *inocybiformis* means "*Inocybe*-shaped," based on the overall resemblance to members of the genus *Inocybe*.

Hygrophorus kauffmanii A. H. Smith and Hesler — Illus. p. 148

SYNONYMS: *Hygrophorus leporinus* Fries

COMMON NAME: none.

CAP: 2–11 cm wide, convex to campanulate or nearly plane; surface canescent, often with minute spot-like scales, viscid to subviscid, chestnut brown to pecan brown when fresh, becoming tawny buff in age; margin incurved and minutely tomentose; flesh light pinkish cinnamon or darker when moist, firm, thick over the center and thin toward the margin; odor and taste not distinctive.

GILLS: adnate to subdecurrent, subdistant to close, broad, pinkish when young, becoming dark reddish brown to orangey cinnamon at maturity; edges even.

STALK: 4–10 cm long, 5–20 mm thick, nearly equal or tapered downward, canescent, solid, dry, glabrous or subsquamulose on the upper portion and appressed-fibrillose below, pinkish buff at first, becoming concolorous with the pileus or darker in age.

MICROSCOPIC FEATURES: spores 7–9 × 4–5.5 µm, ellipsoid, smooth, hyaline, inamyloid.

OCCURRENCE: gregarious to scattered in open woods under oaks; October–February; Michigan and Tennessee; uncommon.

EDIBILITY: unknown.

OBSERVATIONS: This generally large, late-season waxcap is easily overlooked when covered by freshly fallen leaves. The species name honors American agaricologist C. H. Kauffman.

Hygrophorus laurae Morgan — Illus. p. 143

SYNONYMS: *Hygrophorus laurae* var. *decipiens* Peck

COMMON NAME: none.

CAP: 2–10 cm wide, convex to nearly plane, umbonate at times; margin even or undulating to crenate; surface glutinous, silky shining under the gluten, reddish brown to pinkish tan or yellowish brown on the disc, paler or whitish toward the margin; flesh firm, thick, white; odor and taste not distinctive.

GILLS: adnate or decurrent with a tooth, subdistant, rarely forked, attenuated, white at first then tinged pinkish; edges even.

STALK: 2.5–8 cm long, 1–2.5 cm thick, nearly equal or tapered downward, solid, viscid or glutinous near the base, apex with scabrous dots, white to creamy buff.

MICROSCOPIC FEATURES: spores 5.5–8 × 3.5–4.5 µm, ellipsoid, smooth, hyaline, inamyloid.

OCCURRENCE: gregarious to clustered on soil and litter in broadleaf and mixed woods; June–December; throughout the Northeast, south to North Carolina and Tennessee; occasional.

EDIBILITY: unknown.

OBSERVATIONS: This species often fruits earlier in the season than most of the larger waxcaps. *Hygrophorus laurae* var. *decipiens* Peck is nearly identical, but its cap is a little darker with a tinge of brown on the disc. We have placed *Hygrophorus laurae* var. *decipiens* in synonymy because of a lack of distinctive differentiating characteristics.

Hygrophorus monticola A. H. Smith and Hesler Illus. p. 139

SYNONYMS: none.

COMMON NAME: Mountain Waxcap.

CAP: 2–8 cm wide, hemispheric to convex, subumbonate at times; surface glabrous, subviscid but soon dry, margin incurved, salmon-colored over a brownish ground color, disc cinnamon buff; flesh thick, firm, white to yellowish; odor of almond extract; taste not distinctive.

GILLS: decurrent, distant to subdistant, broad, many forked halfway to the margin, intervenose, at times almost poroid, white to cream at first, then flushed the color of the cap.

STALK: 5–9 cm long, 5–20 mm thick, solid, nearly equal or tapered downward, white to pale yellowish on the upper portion, tan-streaked on the lower, sometimes flushed the color of the cap in age.

MICROSCOPIC FEATURES: spores 10–14 × 5.5–7.5 µm, ellipsoid, smooth, hyaline, inamyloid.

OCCURRENCE: gregarious on soil or among mosses under conifers, especially spruce; August–October; Maine, Nova Scotia, and Quebec, also reported from western North America; distribution limits and frequency yet to be determined.

EDIBILITY: edible according to McNeil (2006).

OBSERVATIONS: *Hygrophorus agathosmus* (p. 55) has a similar odor but differs in having a gray, glutinous to viscid cap. The name *monticola* means "mountain dwelling."

Hygrophorus obconicus Peck Not Illustrated

SYNONYMS: *Camarophyllus obconicus* (Peck) Murrill

COMMON NAME: none.

CAP: 1–2.5 cm wide, convex; margin even or sometimes lobed; surface moist, becoming dry, hygrophanous, pruinose to canescent, buff, fading to whitish; flesh brittle, waxy, thick at the center, thin toward the margin, white; odor not distinctive; taste slightly sour or not distinctive.

GILLS: adnate to subdecurrent, close to subdistant, broad, intervenose, white; edges even.

STALK: 2–5 cm long, 2–7 mm thick, nearly equal, hollow, compressed, dry, glabrous, concolorous with cap.

MICROSCOPIC FEATURES: spores 4–5.5 × 3–5 µm, subglobose to short ellipsoid, smooth, hyaline, inamyloid.

OCCURRENCE: on soil in mixed woods; July–September; New England to Tennessee; uncommon.

EDIBILITY: unknown.

OBSERVATIONS: Similar to *Hygrophorus cremicolor* (Murrill) Murrill, which has yellow gills, and *Hygrocybe albipes* (Peck) comb. nov., which has narrow gills, a strongly decurved cap margin, and larger spores that measure 5.5–8 × 4.5–6 µm. The name *obconicus* means "inverted cone."

Hygrophorus olivaceoalbus var. ***olivaceoalbus*** (Fries) Fries Illus. p. 150

SYNONYMS: *Hygrophorus korhonenii* Harmaja

COMMON NAME: Olive Waxcap.

CAP: 3–6.5 cm wide, campanulate to convex becoming nearly plane, often with an umbo; surface glutinous to viscid, drying with conspicuous appressed dark fibrils, center dark brown to black and ashy gray at the margin; flesh thick at the center, soft, white, sometimes yellow under the cuticle; odor and taste not distinctive.

GILLS: adnate to subdecurrent, subdistant, thick, broad, at times intervenose or forked near the margin, white to cream or sometimes with pale gray tinges; edges even.

STALK: 8–15 cm long, 1–3 cm thick, solid, nearly equal overall tapering downward or enlarged downward, covered from the base nearly to the apex by a double sheath; outer sheath glutinous; inner sheath composed of appressed blackish fibers that break into ragged zones encircling the stalk, often with an evanescent glutinous annular zone, glabrous to slightly pruinose above the annular zone; apex and flesh white.

MICROSCOPIC FEATURES: spores 9–12 × 5–8 µm, ellipsoid, smooth, hyaline, inamyloid.

OCCURRENCE: gregarious to scattered among mosses or on coniferous humus, under spruce or hemlock; August–October; Nova Scotia to Michigan and western North America; uncommon in eastern North America.

EDIBILITY: reported to be edible.

OBSERVATIONS: *Hygrophorus olivaceoalbus* var. *gracilis* Maire, recorded from Michigan, Oregon, and Washington, is nearly identical except for having a much thinner stalk that measures 5–7 mm thick. Hesler and Smith (1963) recognized several varieties from western North America. The name *olivaceoalbus* refers to the olivaceus and white colors of this mushroom.

Hygrophorus paludosus Peck

Illus. p. 144

SYNONYMS: none.

COMMON NAME: none.

CAP: 4–10 cm wide, broadly convex to nearly plane, may be depressed in the center at maturity; surface glutinous, the gluten hyaline becoming grayish, cuticle streaked or zoned under the gluten, pale pinkish buff, fibrillose at the margin; flesh thick, firm, white; odor and taste not distinctive.

GILLS: adnate to subdecurrent, close to subdistant, broad, white, usually greenish spotted at maturity.

STALK: 5–12 cm long, 1–2 cm thick, glutinous, nearly equal or tapered downward, base narrowed and subradicating at times, apex scabrous, with punctations that usually become dull yellowish to greenish, whitish below the apex but appearing gray from the gluten, which when dry leaves grayish bands over the lower portion, with a white fibrillose veil under the gluten.

MICROSCOPIC FEATURES: spores 8–11 × 5–7 µm, ellipsoid, smooth, hyaline, inamyloid.

OCCURRENCE: scattered to gregarious in humus in mixed woods; August–January; widely distributed in eastern North America; locally abundant.

EDIBILITY: unknown.

OBSERVATIONS: The greenish spots and punctations that sometimes develop on the apex of the stalk and gills may not be evident in dry weather. The name *paludosus* means "of swamps or marshy ground."

Hygrophorus piceae Kühner

Illus. p. 130

SYNONYMS: none.

COMMON NAME: Spruce Waxcap.

CAP: 1–5 cm wide, convex, becoming nearly plane; margin incurved and remaining so well into maturity, uplifted in age; surface viscid, glabrous to innately appressed-fibrillose, white or at times cream-colored over the center; flesh soft, thick at the center, thin at the margin, white; odor and taste not distinctive.

GILLS: adnate to subdecurrent, subdistant, thin, some forking near the margin, snow white to cream-colored, becoming pinkish buff in age; edges even.

STALK: 3–6 cm long, 8–12 mm thick, nearly equal overall, solid, becoming hollow, apex scaly to fibrillose, glabrous below, dry or slightly tacky, lacking a partial veil and ring, white.

MICROSCOPIC FEATURES: spores 6–8 × 4–6 µm, ellipsoid, smooth, hyaline, inamyloid.

OCCURRENCE: solitary to gregarious under spruce in northern forests; August–November; Maine to Michigan, also reported from California, Idaho, and the Pacific Northwest; fairly common.

EDIBILITY: unknown.

OBSERVATIONS: This northern species is similar to several white waxcaps, including *Hygrocybe virginea* var. *virginea* (p. 54), which lacks the scaly to fibrillose apex on its stalk and has gills that are more strongly decurrent. The name *piceae* refers to the association with spruce trees (genus *Picea*).

Hygrophorus ponderatus Britzelmayr — Not Illustrated

SYNONYMS: none.

COMMON NAME: Ponderous Waxcap.

CAP: 4–12 cm wide, convex when young, becoming broadly convex to nearly plane in age; margin incurved at first and remaining so well into maturity, often wavy or irregular in age; surface viscid when fresh, smooth, white; flesh thick, firm, white, not staining when exposed; odor and taste not distinctive.

GILLS: decurrent, moderately close, white, not staining when bruised.

STALK: 3–7 cm long, 1.2–2 cm thick, nearly equal overall or slightly enlarged at the base, viscid when fresh, solid, smooth, white; with a white, cortina-like, partial veil present on very young stages that soon disappears.

MICROSCOPIC FEATURES: spores 6.5–10 × 5–6 µm, ellipsoid, smooth, hyaline, inamyloid.

OCCURRENCE: solitary, scattered, or in groups on the ground under oak and pine; known from Tennessee, Georgia, and Alabama; September–January; occasional.

EDIBILITY: edible.

OBSERVATIONS: *Hygrophorus sordidus* (p. 76) is nearly identical, but its stalk base is tapered downward; it has narrower spores that measure 6–9 × 3–5 µm; and it lacks a cortina-like partial veil. *Hygrophorus subsordidus* (p. 79) also has a tapered stalk base, has narrower spores that measure 5.5–8 × 3–4 µm, and lacks a cortina-like partial veil. The name *ponderatus* means "large" or "weighty."

Hygrophorus pudorinus var. ***pudorinus*** f. ***pudorinus*** (Fries) Fries — Illus. p. 145

SYNONYMS: none.

COMMON NAME: Blushing Hygrophorus, Turpentine Waxcap.

CAP: 4–11 cm wide, obtuse to convex, becoming broadly convex in age; margin incurved when young, elevated and sometimes uplifted in age; surface smooth but often becoming cracked, especially on the disc, viscid when moist, pinkish buff to pale flesh-colored, pale pinkish salmon, or pale tan; flesh thick, white or tinged pinkish red; odor fragrant, sometimes of turpentine, or not distinctive; taste turpentine-like or not distinctive.

GILLS: subdecurrent, acuminate, subdistant, narrow, sometimes forked and often intervenose, white to whitish, sometimes with pinkish to reddish or salmon tinges.

STALK: 4–9 cm long, 1–2 cm thick, nearly equal or tapered in either direction, dry or somewhat lubricous, rarely slightly viscid, solid, white to buff, sometimes tinged pinkish red, with tiny white scales or points near the apex that quickly stain yellow in KOH.

MICROSCOPIC FEATURES: spores 6.5–10 × 4–5.5 µm, ellipsoid, smooth, hyaline.

OCCURRENCE: scattered or in groups on the ground in conifer woods, especially with spruce, or among sphagnum mosses in bogs; August–October; widely distributed in the Northeast; occasional.

EDIBILITY: edible by most accounts but not recommended.

OSERVATIONS: At least four varieties and forms of this mushroom are recognized and are distinguished primarily by cap color, stalk base color, and cap flesh odor. *Hygrophorus pudorinus* var. *fragrans* (Murrill) Hesler and A. H. Smith has an apricot orange to salmon disc, salmon-colored cap flesh that has a faintly fragrant odor, and a stalk base that is ochraceous orange to orange both inside and out. *Hygrophorus pudorinus* var. *fragrans* f. *pallidus* (A. H. Smith and Hesler) Hesler and A. H. Smith has a white to whitish cap overall or a yellowish tint on its disc, white cap flesh that changes to a pale pink on exposure and that has a slightly fragrant odor, and a stalk base that is white both inside and out. *Hygrophorus pudorinus* var. *subcinereus* (A. H. Smith and Hesler) Hesler and A. H. Smith has a whitish cap that sometimes has a grayish tint over the disc, white unchanging flesh that has a faintly fragrant odor, and a stalk base that is white both inside and out. The name *pudorinus* means "bashful" or "blushing."

Hygrophorus purpurascens (Albertini and Schweinitz) Fries Illus. p. 157

SYNONYMS: none.

COMMON NAME: Veiled Waxcap.

CAP: 3–12 cm wide, convex, becoming broadly convex to nearly flat; surface dry, covered with minute fibrils giving a streaked to somewhat scaly appearance, purplish to dark purplish brown, red-brown, or cinnamon on the disc, pale pink to pinkish tan or whitish toward the margin; margin inrolled when young, incurved to expanded in age, fibrous-cottony; flesh white, thick; odor and taste not distinctive.

GILLS: adnate and often decurrent, subdistant, narrow, white when young, becoming pinkish buff, often spotted or streaked purplish to red-brown, edges even, at times reddish.

STALK: 3–10 cm long, 1–2.5 cm thick, nearly equal, tapered at the base, solid, dry, silky and white at the apex, scurfy and colored like the cap below, with vinaceous brown spots and streaks, staining yellow to yellow-brown in KOH; partial veil white, cottony to fibrillose, leaving a sparse, fugacious, fibrillose ring near the apex, which becomes purplish red at maturity.

MICROSCOPIC FEATURES: spores 5.5–8 × 3–4.5 µm, ellipsoid, smooth, hyaline, inamyloid.

OCCURRENCE: in groups or clusters on the ground under conifers, especially spruce and pine, and in sphagnum bogs; July–October; widely distributed in the Northeast; infrequent.

EDIBILITY: edible but sometimes bitter.

OBSERVATIONS: Young specimens that exhibit evidence of a partial veil are easy to distinguish from otherwise similar species. Compare with *Hygrophorus capreolarius* (p. 58), which has a darker-colored cap and lacks a partial veil. Compare also with *Hygrophorus russula* (p. 76) and *Hygrophorus erubescens* var. *erubescens* (p. 62). The name *purpurascens* means "becoming purple."

Hygrophorus pusillus Peck — Not Illustrated

SYNONYMS: none.

COMMON NAME: none.

CAP: 1–4 cm wide, convex to nearly plane in age; margin incurved at first, becoming uplifted; surface viscid, glabrous, white, center may be tinged yellow or tan; flesh soft, thin, white; odor mildly aromatic, fruity or anise-like when fresh; taste not distinctive.

GILLS: adnate to subdecurrent, subdistant, forked near the margin, white.

STALK: 2–7 cm long, 1–8 mm thick, solid, equal or tapered downward, glabrous, apex sometimes slightly pruinose, white.

MICROSCOPIC FEATURES: spores 7–9 × 4–5 µm, ellipsoid, smooth, hyaline, inamyloid.

OCCURRENCE: scattered to gregarious under conifers; September–December; known mostly from western North America, but also collected in New York; uncommon or possibly rare in eastern North America.

EDIBILITY: unknown.

OBSERVATIONS: The aromatic odor of this species when fresh helps separate it from *Hygrophorus piceae* (p. 71) and *Hygrocybe virginea* var. *virginea* (p. 54). The name *pusillus* means "very small."

Hygrophorus pustulatus (Persoon) Fries — Illus. p. 151

SYNONYMS: *Hygrophorus tephroleucus* var. *tephroleucus* (Persoon) Fries

COMMON NAME: Pepper-Stalk Waxcap, Spotted-Stalk Waxcap.

CAP: 2–5 cm wide, convex to nearly plane, often with a low umbo; surface somewhat viscid to glutinous, fibrillose-scaly, streaked at times, gray to fuscous, grayish brown, or brown over the center, paler toward the margin, at times buff all over; margin incurved at first and remaining so well into maturity; flesh white, soft, thin; odor and taste not distinctive.

GILLS: adnate to subdecurrent, close to subdistant, thick, narrow, intervenose at times, white or tinged pale pinkish; edges even or slightly undulating.

STALK: 6–9 cm long, 5–8 mm thick, nearly equal or slightly enlarged downward, solid, viscid below the apex from remains of a thin, gelatinous universal veil; apex dry, whitish to colored like the cap or paler, with fine dark gray to brown punctations overall.

MICROSCOPIC FEATURES: spores 7–10 × 4–5 µm, ellipsoid, smooth, hyaline, inamyloid.

OCCURRENCE: gregarious among mosses or on soil under conifers, usually with spruce or fir; August–November; Nova Scotia, Ontario, Quebec, Michigan, and West Virginia, but also reported from Colorado, Idaho, Wyoming, and the Pacific Northwest; uncommon.

EDIBILITY: reported as edible by McNeil (2006).

OBSERVATIONS: The dark stalk punctations may disappear in age or in rainy weather. This species resembles *Hygrophorus agathosmus* (p. 55), which can be distinguished by its odor of cherry pits or almond extract. Some authors consider *Hygrophorus tephroleucus* var. *tephroleucus* to be a distinct species, but we cannot find enough distinguishing characters to support this conclusion. The term *pustulatus* means "covered with blisters," in reference to the stalk ornamentation.

Hygrophorus roseibrunneus Murrill Illus. p. 141

SYNONYMS: none.

COMMON NAME: Rosy-Brown Waxcap.

CAP: 2–9 cm wide, convex to broadly convex, becoming nearly plane at maturity, typically with a broad, low central umbo; margin incurved at first, becoming uplifted and wavy in age; surface agglutinated-fibrillose, especially near the margin, viscid when wet but soon dry, dark pinkish brown to pinkish cinnamon, paler and often with rosy pink tints toward the margin at maturity, paler overall in age; flesh thick, soft, white; odor and taste not distinctive.

GILLS: adnate or slightly decurrent, close to crowded, white to very pale yellow; edges even.

STALK: 3–12 cm long, 6–20 mm thick, nearly equal or tapered downward near the base, usually curved, dry, solid, densely pruinose to scurfy, at least on the upper half, white.

MICROSCOPIC FEATURES: spores 6–9 × 3.5–5 µm, ellipsoid, smooth, hyaline, inamyloid.

OCCURRENCE: scattered or in groups on the ground under conifers and broadleaf trees, especially oak; September–March; widely distributed; occasional.

EDIBILITY: reported to be edible.

OBSERVATIONS: A thin, whitish partial veil may be evident on very young specimens, but it soon disappears as the mushroom develops. The name *roseibrunneus* means "rose brown."

Hygrophorus russula (Schaeffer : Fries) Quélet Illus. p. 160

SYNONYMS: *Tricholoma russula* (Fries) Gillet

COMMON NAME: Russula-like Waxcap.

CAP: 5–12 cm wide, hemispheric to convex, becoming broadly convex to nearly plane or depressed in age; margin finely pubescent, white and inrolled when young and remaining so well into maturity; surface viscid at first but soon dry, coated with tiny appressed fibrils and scales, color variable, spotted and streaked vinaceous pink to purplish red over a pale pink to white ground color; flesh thick, firm, white or tinged pinkish, sometimes staining weakly yellow when bruised; odor not distinctive; taste sometimes mild but usually bitter.

GILLS: adnate at first, becoming decurrent at maturity, close to crowded, white when young, soon developing purplish red spots and flushed purplish red in age.

STALK: 3–8 cm long, 1.5–3.5 cm thick, nearly equal or sometimes tapered downward, dry, solid, typically pruinose at the apex and smooth below it, white at first, soon colored like the cap.

MICROSCOPIC FEATURES: spores 6–8 × 3–5 µm, ellipsoid, smooth, hyaline, inamyloid.

OCCURRENCE: solitary, scattered, or in groups on the ground under oaks; August–December; widely distributed; occasional to fairly common.

EDIBILITY: edible but often bitter and unpalatable.

OBSERVATIONS: This waxcap typically has a compact, squatty stature. *Hygrophorus russuliformis* Murrill, which is known only from Florida, differs in having larger spores (8–10 × 2–3.2 µm), and, according to the original description, dried specimens exhibit a sweet, candy-like odor. *Hygrophorus erubescens* var. *erubescens* (p. 62) is a similar and more northern species that is typically less robust, has subdistant to closely spaced gills, grows in conifer woods, and has flesh that stains yellow when exposed and rubbed. Compare with *Hygrophorus purpurascens* (p. 73), which has a partial veil when young and grows in conifer woods. The species epithet refers to the similar aspect to mushrooms in the genus *Russula*.

Hygrophorus sordidus Peck Illus. p. 132

SYNONYMS: none.

COMMON NAME: Sordid Waxcap.

CAP: 5–18 cm wide, convex to nearly plane; surface viscid, glabrous except for a subfloccose margin, white or with a yellowish center; flesh thick, white; odor not distinctive; taste mild to slightly bitter.

GILLS: adnate to subdecurrent, subdistant, intervenose at times, broad, white, becoming yellowish in age; edges even.

STALK: 5–10 cm long, 1.5–2 cm thick, nearly equal overall, often bluntly pointed at the base, solid, dry, glabrous or faintly cottony at the apex, white.

MICROSCOPIC FEATURES: spores 6–9 × 3–5 µm, ellipsoid, smooth, hyaline, inamyloid.

OCCURRENCE: solitary to gregarious on soil in open broadleaf woods, often under oaks; July–November; North Carolina to Canada and west to Illinois, also recorded from California; common.

EDIBILITY: edible.

OBSERVATIONS: *Hygrophorus subsordidus* (p. 79) is similar but has narrower gills, slightly narrower spores, and a more southern distribution, where it typically fruits from November to February. *Hygrophorus ponderatus* (p. 72) is another large white waxcap that is nearly identical, but it has a viscid stalk and an indistinct partial veil on young specimens. The name *sordidus* means "dirty," in reference to the abundant soil, sand, and other debris that adheres to the cap.

Hygrophorus speciosus var. ***speciosus*** Peck — Illus. p. 155

SYNONYMS:

Hygrophorus aureus Arrhenius
Hygrophorus hypothejus var. *aureus* (Arrhenius) Imler
Hygrophorus speciosus var. *kauffmanii* Hesler and A. H. Smith

COMMON NAME: Larch Waxcap.

CAP: 2–5 cm wide, convex to broadly convex, becoming nearly flat, sometimes with an acute umbo; surface smooth, viscid, bright orange-red overall when young, becoming orange to orange-yellow overall or toward the margin or streaked with these colors, typically remaining orange-red over the disc or umbo well into maturity; flesh white to pale yellow; odor and taste not distinctive.

GILLS: adnate to slightly decurrent, sometimes unevenly attached to the stalk, distant to subdistant, moderately broad, waxy, white to yellowish.

STALK: 4–10 cm long, 4–10 mm wide, nearly equal, sometimes enlarged at the base, stuffed, glabrous to fibrillose, viscid when young, whitish to pale amber yellow, often stained tawny orange on the upper portion or overall as the gluten dries, white at the apex above the glutinous covering.

MICROSCOPIC FEATURES: spores 8–10 × 4.5–6 µm, ellipsoid, smooth, hyaline, inamyloid.

OCCURRENCE: scattered, in groups, or in clusters on the ground or among sphagnum mosses under cedar, larch, pine, or spruce; August–November; widely distributed in the northeast; occasional.

EDIBILITY: reported to be edible.

OBSERVATIONS: *Hygrophorus speciosus* var. *kauffmanii* is more robust with a wider cap and thicker stalk. We have placed it in synonymy because of a lack of significantly differentiating characteristics. The name *speciosus* means "showy."

Hygrophorus subfuscescens var. ***subfuscescens*** A. H. Smith and Hesler

Not Illustrated

SYNONYMS: none.

COMMON NAME: none.

CAP: 0.6–2.5 cm broad, convex to broadly convex; margin lobed or crenate; surface glabrous, hygrophanous, yellow, fading to clay or pinkish buff, slowly becoming brown, at times with a gray cast; flesh brittle, thin at the margin and thick at the center, concolorous with the cap; odor mild; taste unpleasant.

GILLS: decurrent, distant, fairly broad, yellowish to whitish at first, soon brown; edges even.

STALK: 2–8 cm long, 1.5–3 mm thick, tapered downward, glabrous, pale yellow, becoming paler in age, but retaining a yellow color.

MICROSCOPIC FEATURES: spores 5–6 × 4–5 µm, subglobose to broadly ellipsoid, smooth, hyaline, inamyloid.

OCCURRENCE: caespitose on bare soil, often in swamps or wet areas, in mixed broadleaf woods, and under hemlock; August–September; reported from Nova Scotia, Maine, and Michigan; uncommon.

EDIBILITY: unknown.

OBSERVATIONS: The northern distribution, habitat, persistently yellow stalk, and changeable cap colors are helpful features for identifying this waxcap. Hesler and Smith (1963) report *Hygrophorus subfuscescens* var. *odora* from Michigan, which has a cap and gills that become gray with age and flesh that has an unpleasant odor when crushed. The name *subfuscescens* means "less than dusky brown."

Hygrophorus subrufescens Peck

Illus. p. 156

SYNONYMS: *Camarophyllus subrufescens* (Peck) Murrill

COMMON NAME: none.

CAP: 1.5–4.5 cm wide, convex to nearly plane; surface scurfy to finely scaly, especially near the margin, pale pink to grayish red; flesh whitish with a pink tinge; odor and taste not distinctive.

GILLS: decurrent, distant, thick, waxy, medium broad, whitish.

STALK: 4–8 cm long, 4–10 mm thick, nearly equal or tapered downward, solid, glabrous, white, streaked pink in age.

MICROSCOPIC FEATURES: spores 5.5–8 × 4–5 µm, ellipsoid, smooth, hyaline, inamyloid.

OCCURRENCE: scattered on the ground in oak woods and under broadleaf trees in mixed woods; July–September; New York and Quebec, distribution limits yet to be determined; uncommon.
EDIBILITY: unknown.
OBSERVATIONS: The name *subrufescens* means "less than reddish."

Hygrophorus subsalmonius A. H. Smith and Hesler Illus. p. 154
SYNONYMS: none.
COMMON NAME: none.
CAP: 6–12 cm wide, convex, becoming broadly convex to flat, depressed on the disc in age; surface smooth, glutinous to viscid, rusty or cinnamon brown when young, fading to pale brown, dull orange, pinkish tan, salmon buff or yellowish white toward the margin in age, disc remaining reddish brown, cinnamon, or shades of orange-brown; margin inrolled when young, with cottony fibers, becoming smooth and somewhat uplifted in age; flesh thick, whitish; odor and taste not distinctive.
GILLS: adnate to decurrent, close to subdistant, waxy, cream to faintly pink at first, becoming ivory to pinkish tan in age.
STALK: 3–10 cm long, 5–20 mm thick, equal or somewhat enlarged at the apex and tapered at the base, solid, smooth, minutely scurfy at the apex, typically slimy at the base from the remnants of a gelatinous universal veil, white to pale tan, staining yellow, orange-yellow, or brownish in age or when bruised.
MICROSCOPIC FEATURES: spores 6.5–8 × 3–5 µm, ellipsoid, smooth, hyaline, inamyloid.
OCCURRENCE: scattered or in groups on the ground under broadleaf trees, especially oak and hickory; August–January; widely distributed; occasional.
EDIBILITY: unknown.
OBSERVATIONS: Soil and debris typically adhere to the glutinous cap on drying. Varieties of *Hygrophorus pudorinus* are similar but less richly colored, and their stalk bases lack the slimy layer of a gelatinous universal veil. The name *subsalmoneus* means "less than salmon-colored."

Hygrophorus subsordidus Murrill Illus. p. 132
SYNONYMS: none.
COMMON NAME: Dirty Southern Waxcap.
CAP: 4–10 cm broad, convex then expanding and often becoming somewhat depressed; surface viscid, glabrous or somewhat finely cracked, white overall or sometimes pale yellow on the disc; margin even, undulate or lobed; flesh thick in the center, thin at the margin, white; odor and taste not distinctive.
GILLS: adnexed or adnate-subdecurrent, narrow, subdistant, some forking midway, white, not staining when handled or bruised; edges even.
STALK: 3–8 cm long, 1–2 cm thick, nearly equal or tapering downward, moist

or dry, usually somewhat lubricous but not viscid, white, pruinose near the apex and fibrillose below it.

MICROSCOPIC FEATURES: spores 5.5–8 × 3–4 µm, oblong-cylindric, smooth, hyaline, inamyloid.

OCCURRENCE: scattered or in groups on the ground in oak–pine woods, oak woods, or grassy areas where oaks are present, especially live oak *(Quercus virginiana);* November–February; reported from Alabama, Florida, and Texas; locally fairly common.

EDIBILITY: edible and good.

OBSERVATIONS: When this waxcap is being collected for eating, it is easier to peel the cuticle than to remove the debris that sticks to the viscid cap. Compare with *Hygrophorus sordidus* (p. 76), which has broader gills, somewhat broader spores, and a more northern distribution and which typically fruits from July to November. The name *subsordidus* means "near *sordidus*," in reference to the likeness to *Hygrophorus sordidus.*

Hygrophorus tennesseensis A. H. Smith and Hesler — Illus. p. 147

SYNONYMS: none.

COMMON NAME: Tennessee Waxcap.

CAP: 5–12.5 cm wide, convex to nearly plane, depressed over the disc at times in age; margin incurved at first; surface smooth, glabrous, viscid when wet, tawny brown over the center, cream to white toward the margin, entire cap pale when young; flesh fairly thick, firm, white; odor of raw potatoes; taste of the flesh bitter, taste of the cap cuticle sour.

GILLS: adnate to subdecurrent, subdistant, intervenose at the margin, white; edges even.

STALK: 4–10 cm long, 1–2 cm thick, nearly equal or tapered downward, solid, dry, scurfy at the apex, fibrillose toward the base, white.

MICROSCOPIC FEATURES: spores 6–10 × 4.5–6 µm, ellipsoid, smooth, hyaline, inamyloid.

OCCURRENCE: gregarious in litter and on soil under conifers, often with hemlock; August–October; New England south to North Carolina and west to Kentucky; fairly common.

EDIBILITY: unknown, but the bitter taste would make it unpalatable.

OBSERVATIONS: The potato-like odor that is often present on fresh specimens helps to distinguish this waxcap from *Hygrophorus pudorinus* var. *pudorinus* f. *pudorinus* (p. 72) and other similar species. The name *tennesseensis* refers to Tennessee, where this waxcap was first described.

COLOR PLATES

For the convenience of users, we have arranged the illustrations according to genus, then in groups within a genus that have similar colors of the fruiting bodies. Because many waxcaps are quite variable and do not always fit cleanly into one color group or another, this arrangement has limitations in the attempt to match specimens with photographs. Nevertheless, we hope it will prove useful as a starting point for identifying most species.

PREVIOUS PAGE: *Hygrocybe miniata* var. *miniata*

Hygrocybe angustifolia

Hygrocybe virginea var. *virginea* (A)

Hygrocybe virginea var. *virginea* (B)

Hygrocybe virginea f. *salmonea*

Hygrocybe russocoriacea

Hygrocybe pratensis var. *pallida* (A)

Hygrocybe pratensis var. *pallida* (B)

Hygrocybe pratensis var. *pallida* (C)

Hygrocybe pratensis var. *pratensis*

Hygrocybe caespitosa (A)

Hygrocybe caespitosa (B)

Hygrocybe hymenocephala (A)

Hygrocybe hymenocephala (B)

Hygrocybe hymenocephala (C)

Hygrocybe deceptiva (A)

Hygrocybe deceptiva (B)

Hygrocybe deceptiva (C)

Hygrocybe nitrata (A)

Hygrocybe nitrata (B)

Hygrocybe fornicata

Hygrocybe colemanniana

Hygrocybe spadicea var. *spadicea*

Hygrocybe irrigata (A)

Hygrocybe irrigata (B)

Hygrocybe ovina

Hygrocybe subovina (A)

Hygrocybe subovina (B)

Hygrocybe subovina (C)

Hygrocybe occidentalis var. *occidentalis*

Hygrocybe canescens (A)

Hygrocybe canescens (B)

Hygrocybe canescens (C)

Hygrocybe canescens (D)

Hygrocybe lacmus (A)

Hygrocybe lacmus (B)

Hygrocybe chlorophana

Hygrocybe parvula

Hygrocybe flavescens (A)

Hygrocybe flavescens (B)

Hygrocybe nitida

Hygrocybe tahquamenonensis

Hygrocybe aurantiosplendens (A)

Hygrocybe aurantiosplendens (B)

Hygrocybe marginata var. *concolor*

Hygrocybe marginata var. *marginata*

Hygrocybe marginata var. *olivacea* (A)

Hygrocybe marginata var. *olivacea* (B)

Hygrocybe auratocephala (A)

Hygrocybe auratocephala (B)

Hygrocybe auratocephala (C)

Hygrocybe purpureofolia

Hygrocybe singeri var. *singeri* (A)

Hygrocybe singeri var. *singeri* (B)

Hygrocybe cantharellus f. *cantharellus* (A)

Hygrocybe cantharellus f. *cantharellus* (B)

Hygrocybe coccineocrenata (A)

Hygrocybe coccineocrenata (B)

Hygrocybe mucronella

Hygrocybe squamulosa (A)

Hygrocybe squamulosa (B)

Hygrocybe squamulosa (C)

Hygrocybe reidii (A)

Hygrocybe reidii (B)

Hygrocybe minutula

Hygrocybe miniata var. *miniata*

Hygrocybe miniata f. *longipes*

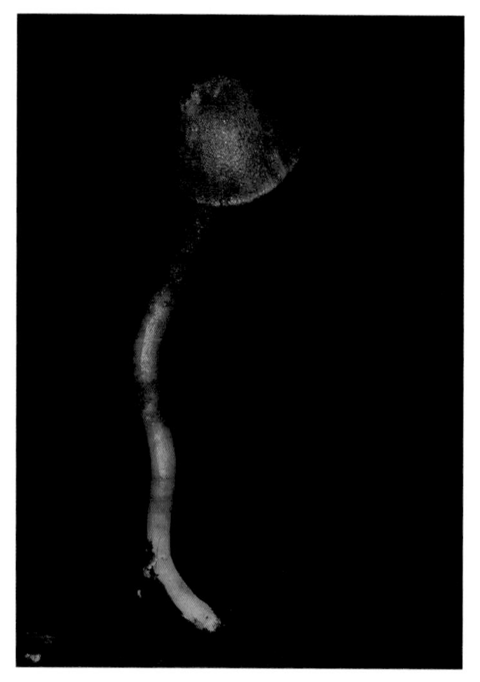

Hygrocybe subminutula

Hygrocybe mississippiensis

Hygrocybe punicea (A)

Hygrocybe punicea (B)

Hygrocybe punicea (C)

Hygrocybe acutoconica var. *acutoconica*

HYGROCYBE CHAMELEONS

Hygrocybe chamaeleon

Hygrocybe appalachianensis

Hygrocybe acutoconica var. *cuspidata* (A)

Hygrocybe acutoconica var. *cuspidata* (B)

Hygrocybe coccinea

Hygrocybe conica var. *conica*

Hygrocybe conica var. *atrosanguinea* (A)

Hygrocybe conica var. *atrosanguinea* (B)

Hygrocybe andersonii (A)

Hygrocybe andersonii (B)

Hygrocybe conicoides (A)

Hygrocybe conicoides (B)

Hygrocybe calyptriformis (A)

Hygrocybe calyptriformis (B)

Hygrocybe laeta var. *laeta* (A)

Hygrocybe laeta var. *laeta* (B)

Hygrocybe laeta var. *laeta* (C)

Hygrocybe laeta var. *laeta* (D)

Hygrocybe psittacina var. *perplexa* (A)

Hygrocybe psittacina var. *perplexa* (B)

Hygrocybe psittacina var. *psittacina* (A)

Hygrocybe psittacina var. *psittacina* (B)

Hygrocybe psittacina var. *psittacina* (C)

HYGROPHORUS

Hygrophorus piceae

Hygrophorus eburneus var. *eburneus* (A)

Hygrophorus eburneus var. *eburneus* (B)

Hygrophorus cossus

Hygrophorus sordidus

Hygrophorus subsordidus (A)

Hygrophorus subsordidus (B)

Hygrophorus subsordidus (C)

Hygrophorus chrysaspis

Hygrophorus chrysodon

Hygrophorus flavodiscus (A)

Hygrophorus flavodiscus (B)

Hygrophorus gliocyclus

Hygrophorus glutinosus (A)

Hygrophorus glutinosus (B)

Hygrophorus glutinosus (C)

Hygrophorus glutinosus (D)

Hygrophorus glutinosus (E)

Hygrophorus monticola

Hygrophorus albofuscus (A)

Hygrophorus albofuscus (B)

Hygrophorus agathosmus (A)

Hygrophorus agathosmus (B)

Hygrophorus roseibrunneus (A)

Hygrophorus roseibrunneus (B)

Hygrophorus roseibrunneus (C)

Hygrophorus roseibrunneus (D)

Hygrophorus laurae (A)

Hygrophorus laurae (B)

Hygrophorus paludosus

Hygrophorus pudorinus var. *pudorinus* f. *pudorinus* (A)

Hygrophorus pudorinus var. *pudorinus* f. *pudorinus* (B)

Hygrophorus discoideus (A)

Hygrophorus discoideus (B)

Hygrophorus bakerensis

Hygrophorus tennesseensis

Hygrophorus kauffmanii

Hygrophorus hypothejus var. *hypothejus*

Hygrophorus camarophyllus

Hygrophorus fuligineus

Hygrophorus olivaceoalbus var. *olivaceoalbus* (A)

Hygrophorus olivaceoalbus var. *olivaceoalbus* (B)

Hygrophorus pustulatus (A)

Hygrophorus pustulatus (B)

Hygrophorus pustulatus (C)

Hygrophorus pustulatus (D)

Hygrophorus inocybiformis

Hygrophorus fuscoalbus var. *fuscoalbus* (A)

Hygrophorus fuscoalbus var. *fuscoalbus* (B)

Hygrophorus subsalmonius (A)

Hygrophorus subsalmonius (B)

Hygrophorus speciosus var. *speciosus* (A)

Hygrophorus speciosus var. *speciosus* (B)

Hygrophorus subrufescens (A)

Hygrophorus subrufescens (B)

Hygrophorus purpurascens (A)

Hygrophorus purpurascens (B)

Hygrophorus capreolarius

Hygrophorus erubescens var. *erubescens* (A)

Hygrophorus erubescens var. *erubescens* (B)

Hygrophorus russula (A)

Hygrophorus russula (B)

Hygrophorus russula (C)

Undescribed Waxcaps

Waxcap species 01

This species has been collected several times during December and January along the Gulf Coast of Florida in Franklin and Gulf counties. It occurs in small groups or caespitose clusters in sandy soil. The distinguishing feature is the deeply rooting stalk with encrusted sand on the lower half or more.

Waxcap species 02

This species was collected during December in the Francis Marion National Forest in South Carolina. It was growing on the ground in mixed oak and pine woods. Although we have collected this medium-size waxcap only once, it may be more commonly encountered and more widely distributed. The distinguishing features are the glutinous infundibuliform pale brown cap with a whitish decurved margin, closely spaced whitish intervenose gills, and a glutinous white stalk.

Undescribed waxcap species 01

Undescribed waxcap species 02

Hygrocybe acutoconica var. *cuspidata*

Glossary

acrid: hot and peppery, sometimes producing a burning sensation in the mouth or throat
adnate (gills): squarely attached to the stalk without a notch
adnexed (gills): notched at the point of attachment to the stalk
agglutinated-fibrillose: having fibrils that are firmly attached as if glued together
annular zone: a poorly defined ring
apex: the uppermost portion of the stalk
apical: pertaining to the uppermost portion of the stalk
apiculus: a short projection at or near the apex of a spore
appressed: flattened onto the surface
appressed-fibrillose: composed of fibrils that are flattened onto the surface
arched: somewhat curved
arcuate: curved like a bow
attenuated: gradually narrowed
base: the lowest portion of the stalk
basidia: club-shaped cells on which spores are formed
bilateral (gill trama): see *divergent*
binucleate: having two nuclei
broadleaf: referring to any tree or shrub with flat leaves
bryophyte: any of numerous small, green, nonvascular plants including mosses, liverworts, and hornworts
caespitose: occurring in groups
calcareous: containing a high proportion of calcium salts
campanulate: bell-shaped
canescent: having a whitish to grayish dust-like bloom
cheilocystidia: cystidia that occur on the edge of a gill
cinereous: gray color tinged with black
clavate: club-shaped
close: spaced halfway between crowded and subdistant
comb. nov.: a new combination
compressed: flattened longitudinally
concolorous: of the same color
conic: shaped more or less like an inverted cone
conical: having the shape of an inverted cone
conic-campanulate: shaped like an inverted cone with a flaring base
conifer: a cone-bearing tree with needle-like leaves
convex: curved or rounded like the exterior of a circle
cortina: a spiderweb-like partial veil
crenate: finely scalloped
crenulate: very finely scalloped
crowded: having little or no space between the gills
cuticle: outermost tissue layer of the cap
cystidia: sterile cells that project between and usually beyond the basidia
decurrent: descending or running down the stalk
decurrent tooth: a hook-like segment between the gill notch and stalk
decurved: bent downward
depressed: sunken
disc: the central area of the surface of a mushroom cap
distant: spaced widely apart
divergent (gill trama): hyphal arrangement typical in *Hygrophorus* where the hyphal elements within the gill diverge obliquely from a more or less parallel central strand; also called *bilateral* (see illus. p. 3)
duff: decaying plant matter on the ground in a forest
ectomycorrhiza: the symbiotic association between plant roots and fungus mycelium whereby the hyphae form a mantle around plant root tips and a net between the plant cells but does not penetrate the cell wall
edges: the thin, outward-facing portion of the gill formed by adjoining faces

ellipsoid: resembling an elongated oval with similarly curved ends
ellipsoid-oblong: resembling an elongated oval with somewhat flattened ends
emarginate: notched near the stalk
entire: even, not broken, serrated, or lacerated
erect: pointing upward or outward
evanescent: slightly developed and soon disappearing
faces: the sides of a gill
farinaceous: having an odor of fresh meal or somewhat resembling a combination of cucumber and watermelon
$FeSO_4$: iron sulfate in water, usually a 10 percent solution
fibrillose: composed of fibrils
fibrillose-pruinose: powdered with tiny fibers
fibrillose-scaly: having fibrils that appear scale-like
fibrillose-scurfy: roughened by partially erect fibrils
fibrillose-squamulose: small scales composed of fibrils
fibrillose-striate: having parallel lines or furrows composed of fibrils
fibrils: tiny fibers
fibrous-cottony: composed of cottony fibers
floccose: tufted like cotton balls
floccose-downy: tufted with very fine fibers
floccose-pruinose: powdered with tiny, somewhat tufted fibers
floccose-pubescent: tufted with hairs
floccose-tomentose: somewhat tufted and matted with hairs
forked (gills): divided to form branches
free (gills): not attached to the stalk
fuscous: dark olive brown to brownish gray or brownish black
fusiform: spindle-shaped and narrowing at both ends
glabrous: bald; lacking hairs, scales, wrinkles, or warts
glandular dots: sticky spots on the surface of the stalk
glaucous: covered with a thin, whitish bloom that is easily rubbed off
gluten: a sticky, glue-like pectinous material
glutinous: having or composed of gluten
granulose: covered with a granule-like substance
gregarious: closely scattered over a small area
humus: decaying organic matter mostly of plant origin
hyaline: transparent; clear and nearly colorless
hygrophanous: appearing water soaked when fresh, fading to a paler color as water is lost
hyphae (sing. **hypha**): threadlike filaments of fungal cells
inamyloid: unchanging or pale yellow in Melzer's reagent
incurved: bent inward toward the stalk
infundibuliform: funnel-shaped
inrolled: bent inward toward the stalk and upward
intervenose: having veins on the gill faces that extend between the gills or from gill to gill
interwoven (gill trama): hyphal arrangement within the gill tissue where the hyphal elements are interlaced with each other but may also include subparallel hyphal cells; also called *irregular gill trama* (see illus. p. 3)
involute: turned inward
KOH: potassium hydroxide in water, usually a 3–5 percent concentration
lacerated: torn or shredded
lacerate-squamulose: torn into tiny scales
lacrymoid: shaped like a teardrop
litter: the uppermost and slightly decayed layer of organic matter on the forest floor
lobed: having rounded divisions
lubricous: smooth and slippery
margin: the edge of a mushroom cap
marginate: having gill edges that are darker colored than the faces
matted-fibrillose: composed of fibrils that are flattened down
Melzer's reagent: a solution containing iodine used for testing spores' color reactions
mephitic: skunk-like odor, foul-smelling, putrid
mycelium: a mass of entangled thread-like filaments
mycorrhizal: having a mutually beneficial relationship with a tree or other plant
nitrous: a reference to nitrogen-containing compounds that are pungent and unpleasant
obovoid: ovoid, with the broader end opposite to the point of attachment
obtuse: rounded or blunt
ochraceous: pale brownish orange-yellow
opaque: dull, not shiny
orbicular: spherical or ball-shaped
ovoid: somewhat egg-shaped
parallel (gill trama): hyphal arrangement

typical in most hygrocybes where the hyphal elements within the gill are elongated and mostly parallel to each other; also called *regular gill trama* (see illus. p. 3)

partial veil: membranous or fibrous material that extends from the edge of the cap to the stalk and covers the immature gills; rupturing and typically leaving velar remnants on the stalk or margin of the cap at maturity as the cap expands

phaseoliform: bean-shaped

plane: flat

pleurocystidia: cystidia that occur on the gill faces

poroid: resembling pores or composed of pores

pruinose: appearing finely powdered

punctate: marked with tiny points, dots, scales, or spots

punctations: tiny points, dots, scales, or spots

pungent: sharp or irritating

rancid: offensive or foul

recurved: curved backward or downward

resinous: tending to adhere

rhizosphere: the region where a plant root and fungal tissue interact

rimose: having distinct cracks or crevices

saprobe: an organism that lives off of dead or decaying matter

saprobic: living off of dead or decaying matter

scabrous: having short, rigid projections

scarlet: deep bright red with an orange or yellow tinge

scurfy: roughened by tiny flakes or scales

seceding: separating away from

serrate: jagged or toothed like a saw blade

serratulate: finely jagged or toothed like a saw blade

sinuate: gradually narrowed and becoming concave near the stalk

spathulate: spoon-shaped

spores: microscopic reproductive cells

squamules: small scales

squamulose: having small scales

striate: having small and more or less parallel lines or furrows

stuffed: containing a soft tissue that usually disappears in age, leaving a hollow space

subdecurrent: extending slightly down the stalk

subdistant: gill spacing halfway between close and distant

subellipsoid: somewhat ellipsoid

subglobose: nearly round

subovate: nearly oval

subovoid: nearly egg-shaped

subradicating: somewhat rooted

subsquamulose: having somewhat small scales

substrate: the nutritional source on which the fungus is growing

subumbonate: having a slight umbo

subviscid: slightly sticky or tacky

superior: located on the upper portion

terete: rounded like a broom handle

tomentose: coated with a thick, matted covering of hairs

tomentum: a coating of soft fibrils

trama: supporting tissue

translucent-striate: appearing to have small lines or furrows when viewed through moist cap tissue

turgor pressure: the pressure exerted on a cell wall by its contents

twisted-striate: having lines or furrows that turn more or less in a helical pattern along the length of the stalk

umbo: a pointed or rounded elevation at the center of a mushroom cap

umbonate: having an umbo

undulate: wave-like

undulating: wavy

uninucleate: having one nucleus

universal veil: a layer of sterile tissue that encloses the entire fruitbody of a developing mushroom; thick and glutinous on many species of *Hygrophorus*

velar: pertaining to a veil

venose: having veins

ventricose: swollen in the middle and tapered to somewhat of a point

vinaceous: pinkish red to pale purplish red

viscid: sticky or tacky

zonate: arranged in concentric bands of different colors or hues

Resources and Recommended Reading

Arora, D. 1986. *Mushrooms Demystified*. Ten Speed Press, Berkeley, CA. 1,020 pp.

Barron, G. 1999. *Mushrooms of Northeast North America*. Lone Pine, Vancouver. 336 pp.

Bas, C., T. H. W. Kuyper, M. E. Noordeloos, and E. C. Vellinga, eds. 1992. *Flora Agaracina Neerlandica 2*. CRC Press, Boca Raton, FL. 137 pp.

Bessette, A. E. 1988. *Mushrooms of the Adirondacks: A Field Guide*. North Country Books, Utica, NY. 145 pp.

Bessette, A. E., A. R. Bessette, and D. W. Fischer. 1997. *Mushrooms of Northeastern North America*. Syracuse Univ. Press, Syracuse, NY. 584 pp.

Bessette, A. E., O. K. Miller Jr., A. R. Bessette, and H. H. Miller. 1995. *Mushrooms of North America in Color: A Field Guide Companion to Seldom-Illustrated Fungi*. Syracuse Univ. Press, Syracuse, NY. 188 pp.

Bessette, A. E., W. C. Roody, A. R. Bessette, and D. L. Dunaway. 2007. *Mushrooms of the Southeastern United States*. Syracuse Univ. Press, Syracuse, NY. 373 pp.

Bessette, A. E., and W. J. Sundberg. 1987. *Mushrooms: A Quick Reference Guide to Mushrooms of North America*. Macmillan, New York. 174 pp.

Bessette, A. R., A. E. Bessette, and W. J. Neill. 2001. *Mushrooms of Cape Cod and the National Seashore*. Syracuse Univ. Press, Syracuse, NY. 174 pp.

Bigelow, H. E. 1959. Interesting Fungi from Massachusetts. *Rhodora* 61: 127–136.

Bigelow, H. E., and M. E. Barr. 1960. Contributions to the Fungus Flora of Northeastern North America. *Rhodora* 62: 186–198.

Binion, D., S. L. Stephenson, W. C. Roody, H. H. Burdsall, L. N. Vasilyeva, and O. K. Miller Jr. 2008. *Macrofungi Associated with Oaks of Eastern North America*. West Virginia Univ. Press, Morgantown. 467 pp.

Bird, C. J., and D. W. Grund. 1979. Nova Scotian Species of *Hygrophorus*. *Proc. Nova Scotia Inst. of Sci.* 29: 1–131.

Boertmann, D. 1995. *Fungi of Northern Europe*. Vol. 1: *The Genus Hygrocybe*. Svampetryk, Greve, Denmark. 184 pp.

Breitenbach, J., and F. Kranzlin. 1991. *Fungi of Switzerland*. Vol. 3: *Boletes and Agarics*. Part 1. Edition Mykologia, Lucerne, Switzerland. 361 pp.

Candusso, M. 1997. *Hygrophorus s.l. Fungi Europaei 6*. Edizioni Candusso, Alassio, Italy. 784 pp.

Cibula, W. G. 1979. Fungi of the Gulf Coast 1: Two New Species of *Hygrophorus* Section *Hygrocybe*. *Mycotaxon* 10 (1): 105–115.

Gourley, C. O. 1983. An Annotated Index of the Fungi of Nova Scotia. *Proc. Nova Scotia Inst. of Sci.* 32: 75–293.

Groves, J. W. 1979. *Edible and Poisonous Mushrooms of Canada*. 2nd rev. ed. Research Branch, Agriculture Canada, Ottawa. 326 pp.

Hansen, L., and H. Knudsen, eds. 1992. *Nordic Macromycetes*. Vol. 2. Nordsvamp, Copenhagen. 474 pp.

Hesler, L. R., and A. H. Smith. 1963. *North American Species of Hygrophorus*. Univ. of Tennessee Press, Knoxville. 416 pp.

Homola, R. L., M. M. Czapowskyj, and B. M. Blum. 1985. The *Ectomycorrhizae of Maine 3: A Listing of Hygrophorus with the Associated Hosts*. Bulletin no. 810. University of Maine, Orono. 19 pp.

Horn, B., R. Kay, and D. Abel. 1993. *A Guide to Kansas Mushrooms*. Univ. Press of Kansas, Lawrence. 297 pp.

Huffman, D. M., L. H. Tiffany, G. Knaphus, and R. A. Healy. 2008. *Mushrooms and Other Fungi of the Midcontinental United States*. 2nd ed. Univ. of Iowa Press, Iowa City. 370 pp.

Kauffman, C. H. 1906. Unreported Michigan Fungi. *Mich. Acad. Sci. Rep.* 8: 26.

———. 1918. *The Agaricaceae of Michigan*. Vol. 1. Wynkoop Hallenbeck Crawford, Lansing, MI. 924 pp.

———. 1922. The Mycological Flora of the Higher Rockies of Colorado. *Papers Mich. Acad. Sci., Arts, and Letters* 1: 101–150.

———. 1926. The Fungus Flora of Mt. Hood with Some New Species. *Papers Mich. Acad. Sci., Arts, and Letters* 5: 115-148.

———. 1971a. *The Gilled Mushrooms (Agaricales) of Michigan and the Great Lakes Region.* Vol. 1. Reprint. Dover, Mineola, NY. 442 pp.

———. 1971b. *The Gilled Mushrooms (Agaricales) of Michigan and the Great Lakes Region.* Vol. 2. Reprint. Dover, Mineola, NY. 481 pp.

Kauffman, C. H., and A. H. Smith. 1933. Agarics Collected in the Vicinity of Rock River, Michigan, in 1929. *Papers Mich. Acad. Sci., Arts, and Letters* 17: 153–200.

Kimbrough, J. 2000. *Common Florida Mushrooms.* Univ. of Florida, Gainesville. 342 pp.

Kirk, P. M., P. F. Cannon, J. C. David, and J. A. Staplers., eds. 2002. *Ainsworth and Bisby's Dictionary of the Fungi.* 9th ed. CABI and Oxford Univ. Press, Oxford. 672 pp.

Largent, D. L. 1985. *Hygrophoraceae.* Agaricales of California no. 5. Mad River Press, Eureka, CA. 208 pp.

Lincoff, G. H. 1981. *The Audubon Society Field Guide to North American Mushrooms.* Knopf, New York. 498 pp.

McIlvaine, Charles. 1900. *Toadstools, Mushrooms, Fungi Edible and Poisonous: One Thousand American Fungi. How to Select and Cook the Edible; How to Distinguish and Avoid the Poisonous.* Bowen-Merrill, Indianapolis, IN. 704 pp.

McKnight, K. H., and V. B. McKnight. 1987. *A Field Guide to Mushrooms.* Houghton Mifflin, Boston. 429 pp.

McNeil, R. 2006. *Le grand livre des champignons du Québec et de l'est du Canada.* Éditions Michel Quintin, Waterloo, Canada. 575 pp.

Metzler, S., V. Metzler, and O. K. Miller Jr. 1992. *Texas Mushrooms.* Univ. of Texas Press, Austin. 350 pp.

Miller, O. K., Jr. 1973. *Mushrooms of North America.* E. P. Dutton, New York. 360 pp.

———. 2006. *North American Mushrooms.* Globe Pequot Press, Guilford, CT. 583 pp.

Murrill, W. A. 1910. Illustrations of Fungi—VII. *Mycologia* 2: 159–163.

———. 1916. *Hydrocybe, Camarophyllus,* and *Hygrophorus. North Amer. Flora* 9: 376–396.

———. 1938. New Florida Agarics. *Mycologia* 30: 359–371.

———. 1939. Some Florida Gill-Fungi. *Elisha Mitchell Sci. Soc. Jour.* 55: 361–372.

———. 1940. Additions to Florida Fungi III. *Torrey Bot. Club Bull.* 67: 145–154.

———. 1941. More Florida Novelties. *Mycologia* 33: 434–448.

———. 1942. New Fungi from Florida. *Lloydia* 5: 136–157.

———. 1943. Some Southern Novelties. *Mycologia* 35: 422–433.

———. 1944. More Fungi from Florida. *Lloydia* 7: 303–327.

———. 1945. New Florida Fungi. *Florida Acad. Sci. Proc.* 7: 107–127.

Peck, C. H. 1907. New York Species of Hygrophorus. *NY State Mus. Bull.* 116: 45–67.

Phillips, R. 1991. *Mushrooms of North America.* Little, Brown, Boston. 319 pp.

Roody, W. C. 2003. *Mushrooms of West Virginia and the Central Appalachians.* Univ. Press of Kentucky, Lexington. 520 pp.

Sicard, M., and Y. Lamoureux. 1999. *Les champignons sauvages du Québec.* Fides, Quebec. 399 pp.

Singer, R. 1951. The "Agaricales" (Mushrooms) in Modern Taxonomy. *Lilloa* 22: 1–768.

Singer, R., and L. R. Hesler. 1939. Studies in North American Species of *Hygrophorus* I: The Subgenus *Limacium. Lloydia* 2: 1–62.

———. 1940. New and Unusual Agarics from the Great Smoky Mountains National Park. *Elisha Mitchell Sci. Soc. Jour.* 56: 302–324.

Smith, A. H. 1938. *Common Edible and Poisonous Mushrooms of Southeastern Michigan.* Cranbrook Institute of Science, Bloomfield Hills, MI. 71 pp.

Smith, A. H., and L. R. Hesler. 1939. Studies in North American Species of *Hygrophorus* I. *Lloydia* 2 (1): 1–62.

———. 1942. Studies in North American Species of *Hygrophorus* II. *Lloydia* 5: 1–94.

———. 1954. Additional North American *Hygrophori. Sydowia* 8: 304–333.

Young, A. M. 2005. *Fungi of Australia: Hygrophoraceae.* ABRS, Canrra, Australia; CSIRO Publishing, Melbourne. 179 pp.

The North American Mycological Association (NAMA) is a nationwide organization that supports and promotes amateur mycology. NAMA holds an annual mushroom foray at various locations around the country, and its Web site (http://www.namyco.org) is a wealth of information with links to regional clubs and activities.

Index to Common Names

Index to Scientific Names

Page numbers in **bold** denote color illustrations.

PHOTO CREDITS

Unless otherwise noted here, photos are by the authors.

DAN GURAVICH
Hygrophorus roseibrunneus (C)

RICHARD HOMOLA
Hygrophorus monticola

EMILY JOHNSON
Hygrophorus subsalmonius (A)

DAVID LARGENT
Hygrocybe subminutula
Hygrophorus discoideus (A)

RENÉE LEBEUF
Hygrocybe angustifolia
Hygrocybe canescens (A, B)
Hygrocybe hymenocephala (A, B, C)
Hygrocybe minutula
Hygrocybe nitrata (A)
Hygrocybe reidii (B)
Hygrocybe tahquamenonensis
Hygrophorus albofuscus (A, B)
Hygrophorus eburneus var. *eburneus* (A)
Hygrophorus laurae (A, B)
Hygrophorus olivaceoalbus var. *olivaceoalbus* (A)
Hygrophorus piceae
Hygrophorus roseibrunneus (A, B)
Hygrophorus subrufescens (A)

DAVID LEWIS
Hygrocybe chamaeleon
Hygrocybe mississippiensis
Hygrocybe virginea f. *salmonea* = *Hygrophorus borealis* f. *salmoneus*

ORSON K. MILLER JR.
Hygrophorus discoideus (B)
Hygrophorus kauffmanii
Hygrophorus paludosus

JOHN PLISCHKE III
Hygrocybe cantharellus f. *cantharellus* (A)
Hygrocybe deceptiva (A)
Hygrocybe flavescens (A)
Hygrocybe irrigata (A)
Hygrocybe lacmus (A)
Hygrocybe laeta var. *laeta* (A)
Hygrocybe pratensis var. *pallida* (A)
Hygrocybe psittacina var. *psittacina* (A)
Hygrocybe singeri var. *singeri* (A)
Hygrocybe squamulosa (A, B)
Hygrophorus flavodiscus (A)
Hygrophorus glutinosus (B)

MICHAEL WOOD
Hygrocybe singeri var. *singeri* (B)

ABOUT THE AUTHORS

Alan E. Bessette is a professional mycologist and emeritus professor of biology at Utica College of Syracuse University. He has published numerous papers in the field of mycology and has authored or coauthored more than twenty books, including *Edible Wild Mushrooms of North America* (1992), *Mushrooms of Northeastern North America* (Syracuse Univ. Press, 1997), and *Mushrooms of the Southeastern United States* (Syracuse Univ. Press, 2007). Alan has presented numerous mycological programs, was the scientific adviser to the Mid-York Mycological Society and served as a consultant for the New York State Poison Control Center for more than twenty years. He has been the principal mycologist at national and regional forays and was the recipient of the 1987 Mycological Foray Service Award and of the 1992 North American Mycological Association Award for Contributions to Amateur Mycology.

William C. Roody is a mycologist and wildlife diversity biologist with the Wildlife Diversity Program at the West Virginia Division of Natural Resources. He has authored or coauthored several books on mushrooms and other macrofungi, including *North American Boletes: A Guide to the Fleshy Pored Mushrooms* (Syracuse Univ. Press, 2000), *Mushrooms of West Virginia and the Central Appalachians* (2003), and *Macrofungi Associated with Oaks in Eastern North America* (2008). He frequently lectures on various aspects of mycology and has taught many mushroom-identification workshops and Master Naturalist classes on fungi. He has won numerous awards in the North American Mycological Association's annual photo competition, including top honors in both the documentary and the pictorial divisions. In 2000, Bill received the North American Mycological Association Award for Contributions to Amateur Mycology.

Walter E. Sturgeon is a nationally recognized mycologist and served as the first vice president of the North American Mycological Association. He was the 1989 recipient of the North American Mycological Association Award for Contributions to Amateur Mycology and the Northeast Mycological Federation Eximia Award in 1993. He has written numerous articles on the identification of wild mushrooms and is the author of "Mushrooms and Other Fungi of the Upland Forests" in *Upland Forests of West Virginia* (1993). Walt also authored the section on mycorrhiza in *Encyclopedia of Appalachia* (2006). His mushroom photographs have won several awards in the North American Mycological Association's annual photo competition, including top honors.

Arleen R. Bessette is a psychologist, mycologist, and botanical photographer who has been collecting and studying wild mushrooms for more than forty years. She has authored or coauthored twelve books, including *Taming the Wild Mushroom: A Culinary Guide to Market Foraging* (1993), *Mushrooms of Northeastern North America* (Syracuse Univ. Press, 1997), *Mushrooms of Cape Cod and the National Seashore* (Syracuse Univ. Press, 2001), and *The Rainbow Beneath My Feet: A Mushroom Dyer's Field Guide* (Syracuse Univ. Press, 2001). Arleen has won several awards in the North American Mycological Association's annual photo competition, including top honors in both the documentary and the pictorial divisions. She teaches introductory courses in mycology, dyeing with mushrooms, and the culinary aspects of mycophagy.